T0332913

High-Temperature Mechanical Hysteresis in Ceramic-Matrix Composites

This book focuses on mechanical hysteresis behavior in different fiber-reinforced ceramic-matrix composites (CMCs), including one-dimensional (1D) minicomposites, 1D unidirectional, 2D cross-ply, 2D plain-woven, 2.5D woven, and 3D needle-punched composites.

CMCs are considered to be the lightweight high-temperature materials for hot-section components in aeroengines with the most potential. To improve the reliability and safety of CMC components during operation, it is necessary to conduct damage and failure mechanism analysis and to develop models to predict this damage, as well as fracture over lifetime – mechanical hysteresis is a key damage behavior in fiber-reinforced CMCs. The appearance of hysteresis is due to a composite's internal damage mechanisms and modes, such as matrix cracking, interface debonding, and fiber failure. Micromechanical damage models and constitutive models are developed to predict mechanical hysteresis in different CMCs. The effects of a composite's constituent properties, stress levels, and the damage states of the mechanical hysteresis behavior of CMCs are also discussed. This book also covers damage mechanisms, damage models, and micromechanical constitutive models for the mechanical hysteresis of CMCs.

This book will be a great resource for students, scholars, material scientists, and engineering designers who would like to understand and master the mechanical hysteresis behavior of fiber-reinforced CMCs.

Longbiao Li is a lecturer at the College of Civil Aviation at the Nanjing University of Aeronautics and Astronautics. Dr. Li's research focuses on the vibration, fatigue, damage, fracture, reliability, safety, and durability of aircraft and aeroengines. In this research area, he is the first author of 188 SCI journal publications, 9 monographs, 4 edited books, 5 textbooks, 3 book chapters, 30 Chinese Patents, 2 US Patents, 2 Chinese Software Copyrights, and more than 30 refereed conference proceedings. He has been involved in different projects related to structural damage, reliability, and airworthiness design for aircraft and areoengines, supported by the Natural Science Foundation of China, COMAC Company, and AECC Commercial Aircraft Engine Company.

High-Temperature Mechanical Hysteresis in Ceramic-Matrix Composites

Longbiao Li

CRC Press
Taylor & Francis Group
Boca Raton London New York

CRC Press is an imprint of the
Taylor & Francis Group, an **informa** business

First edition published 2023
by CRC Press
6000 Broken Sound Parkway NW, Suite 300, Boca Raton, FL 33487-2742

and by CRC Press
4 Park Square, Milton Park, Abingdon, Oxon, OX14 4RN

CRC Press is an imprint of Taylor & Francis Group, LLC

© 2023 Longbiao Li

Library of Congress Cataloging-in-Publication Data
Names: Li, Longbiao, 1983- author.
Title: High-temperature mechanical hysteresis in ceramic-matrix composites / Longbiao Li.
Description: First edition. | Boca Paton : CRC Press, 2023. | Includes bibliographical references and index. |
Summary: "This book focuses on mechanical hysteresis behavior in different fiber-reinforced ceramic-matrix composites (CMCs), including 1D minicomposites, 1D unidirectional, 2D cross-ply, 2D plain-woven, 2.5D woven, and 3D needle-punched composites. Ceramic-matrix composites (CMCs) are considered to be the lightweight high-temperature materials for hot-section components in aeroengines with the most potential. To improve the reliability and safety of CMC components during operation, it is necessary to conduct damage and failure mechanism analysis, and to develop models to predict this damage as well as fracture over lifetime - mechanical hysteresis is a key damage behavior in fiber-reinforced CMCs. The appearance of hysteresis is due to a composite's internal damage mechanisms and modes, such as, matrix cracking, interface debonding, and fiber failure. Micromechanical damage models and constitutive models are developed to predict mechanical hysteresis in different CMCs. Effects of a composite's constituent properties, stress level, and the damage states of the mechanical hysteresis behavior of CMCs are also discussed. This book also covers damage mechanisms, damage models and micromechanical constitutive models for the mechanical hysteresis of CMCs. This book will be a great resource for students, scholars, material scientists and engineering designers who would like to understand and master the mechanical hysteresis behavior of fiber-reinforced CMCs"-- Provided by publisher.
Identifiers: LCCN 2022007125 (print) | LCCN 2022007126 (ebook) | ISBN 9781032316109 (hardback) | ISBN 9781032316154 (paperback) | ISBN 9781003310570 (ebook)
Subjects: LCSH: Ceramic-matrix composites. | Hysteresis. | Heat resistant materials. | Fiber-reinforced ceramics.
Classification: LCC TA418.9.C6 L529 2023 (print) | LCC TA418.9.C6 (ebook) | DDC 620.1/18--dc23/eng/20220504
LC record available at https://lccn.loc.gov/2022007125
LC ebook record available at https://lccn.loc.gov/2022007126

ISBN: 978-1-032-31610-9 (hbk)
ISBN: 978-1-032-31615-4 (pbk)
ISBN: 978-1-003-31057-0 (ebk)

DOI: 10.1201/b23026

Typeset in Minion Pro
by SPi Technologies India Pvt Ltd (Straive)

To My Son Shengning Li

Content

Preface

Advanced high-thrust-to-weight ratio aeroengines need to reduce structural mass and increase the inlet turbine temperature. At present, the performance of superalloy materials is close to the limit, and it is difficult to meet the needs of future aeroengines. Compared with superalloy, the density of fiber-reinforced ceramic-matrix composites (CMCs) is only approximately 1/3 of that of the superalloy, and the operating temperature can reach approximately 1350°C for long-term use. Therefore, CMCs are considered the most potential lightweight high-temperature materials for hot-section components in aeroengines. The SNECMA company has carried out the research on the application of CMCs in the tail nozzles of aeroengines and developed and tested C/SiC and SiC/SiC as the tail nozzle outer and inner adjusting plate of the M88-2 aeroengine, respectively. The strength of C/SiC composite is high, which can be maintained at 250 MPa at 700°C. It has been used in M88 series aeroengines and can meet the long-term requirements of aeroengines. The GE company prepared a full-scale SiC/SiC combustion liner using the slurry cast melting infiltration method and performed the ground test. To improve the reliability and safety of CMCs components during operation, it is necessary to perform investigations on damage and failure mechanisms analysis and develop models to predict the damage, fracture, and lifetime.

Mechanical hysteresis behavior is a key damage behavior in fiber-reinforced CMCs. The appearance of the hysteresis is due to s composite's internal damage mechanisms and modes, such as matrix cracking, interface debonding, and fiber failure. This book focus on the mechanical hysteresis behavior in different fiber-reinforced CMCs, including, one-dimensional (1D) minicomposite, 1D unidirectional, 2D cross-ply, 2D plain-woven, 2.5D woven, and 3D needle-punched. An *in situ* experimental technique is used for the damage evolution in CMCs under cyclic loading. Micromechanical damage models and constitutive models are developed to predict the mechanical hysteresis behavior in different CMCs. The effects of a composite's constituent properties, stress level, and damage states on the mechanical hysteresis behavior of CMCs are also discussed.

This book covers the damage mechanisms, damage models, and micromechanical constitutive models for mechanical hysteresis of CMCs. I hope this book can help material scientists and engineering designers with understanding and mastering the mechanical hysteresis behavior of fiber-reinforced CMCs.

Longbiao Li

CHAPTER 1

Introduction

1.1 APPLICATION BACKGROUND OF CERAMIC-MATRIX COMPOSITES ON AIRCRAFT OR AEROENGINE

The development of a new generation of lightweight, high-efficiency aeroengine requires ceramic-matrix composite (CMC) materials to gradually replace superalloys in hot-section components [1, 2]. Since the 1950s, European and US researchers have carried out research on the application of CMC in aeroengine hot-section components. Among them, SNECMA company and GE company started the research in this field first, and the technology maturity and application are relatively high. Since 1979, the US has made a lot of research and development investments in the application of CMC materials in aeroengines and carried out programs such as High Temperature Engine Materials Technology Program (HITEMP), HighSpeed Research (HSR)–Enabling Propulsion Research (EPM), Ultra-Efficient Engine Technology (UEET), and Environmentally Responsible Aviation (ERA). Other countries also actively carry out research on the application of CMC materials in aeroengines, mainly including the Japanese Advanced Materials Gas-Generator (AMG) program. Finally, a complete set of research systems of CMC material mechanical behavior and its application in aeroengine components has been established, which

DOI: 10.1201/b23026-1

TABLE 1.1   Application of CMCs in Military and Commercial Aeroengines

Aeroengine	Country	Fiber/Matrix	Aircraft	Application Component
M88-2	France	SiC/SiC	Dassaut Rafale	Seals
M52-2	France	C/SiC	Mirage 2000	Exhaust nozzle seal
F119	USA	C/SiC	F22	Vector nozzle inner/outer wall
F414	USA	SiC/SiC	F/A-18	Combustor
F100	USA	SiC/SiC	F-15/F-16	Exhaust nozzle seal
F110	USA	SiC/C and SiC/SiC	F-15, F/A-18	Seals
CFM56	USA/France	SiC/SiC	Boeing 787	Exhaust nozzle
Trent 1000	USA/UK	SiC/SiC	Boeing 787	Combustor liner
Leap-1A	USA/France	SiC/SiC	A320neo	High pressure turbine shroud ring
GE-9X	USA	SiC/SiC	Boeing 777X	Combustor Turbine guide vane
Passport 20	UAS	Nextel 720/Al_2O_3	Global 7000/8000	Mixer/Center body/Engine core cowl

directly promotes the rapid development of aeroengine. Table 1.1 listed the main application of CMCs in military and commercial aeroengines.

1.2 MANUFACTURING OF CMCS

CMC material is composed of high-strength carbon or ceramic fiber and ceramic matrix. On the basis of inheriting the advantages of high-temperature resistance of monolithic ceramic, the purpose of increasing the toughness of the material is achieved through the design of toughening mechanism. CMC materials generally include four structural units: reinforcing fiber, ceramic matrix, interphase between reinforcing fiber and ceramic matrix, and surface environmental barrier coating.

Taking SiC/SiC CMC as an example, the strength and brittleness are greatly improved by introducing SiC reinforcing fiber into the SiC ceramic matrix. The SiC/SiC CMC retains the advantages of SiC ceramic, such as high-temperature resistance, oxidation resistance, wear resistance, low density, and corrosion resistance. At the same time, the strengthening and toughening mechanism of SiC fiber makes the material insensitive to cracks, which overcomes the fatal weaknesses of ceramic materials, such as high brittleness and poor reliability. Compared with nickel-based superalloys, CMC materials have the following significant advantages:

- They can withstand higher temperatures than superalloys (the temperature resistance limit of CMC material is about 150 K higher than that of nickel-based superalloy, up to 1500 K), which can significantly reduce the consumption of cooling air by about 15–25% so as to improve the aeroengine efficiency.

- The material density of CMC (2.0–2.5 g/cm^3) is 1/4–1/3 that of superalloys, which can significantly reduce the aeroengine weight (30–70% of the aeroengine weight reduction), so as to greatly improve the thrust–weight ratio.

- CMC materials have excellent endurance strength at high temperatures.

- They have strong designability. The introduction of fiber textile technology has greatly improved the designability and structural adaptability of CMC. With the development of computer-aided engineering technology, the CMC material mechanics analysis model has also developed to multiscale, and the coupling between material design and structural design has been improved. It can be designed according to the performance requirements of different components so as to achieve the best thermal/mechanical characteristic matching.

1.2.1 Fibers

The reinforcing fibers of CMCs mainly include non-oxide continuous fibers, such as carbon fibers and silicon carbide fibers. For oxide matrix composites, reinforcing fibers include oxide fibers and non-oxide fibers.

1.2.1.1 Carbon Fiber

Carbon fiber can be divided into rayon-based, polyacrylonitrile (PAN)-based, pitch-based, and vapor-grown carbon fiber according to different raw materials. Compared with ceramic fiber, the remarkable advantage of carbon fiber is that it can maintain stable mechanical properties in a large temperature range. The disadvantage of carbon fiber is that oxidation will occur in air above 400°C.

According to the mechanical properties of carbon fibers, they are classified into high-strength, medium modulus, medium modulus high-strength, and high modulus carbon fibers. High-strength carbon fibers possess a strength of 2.0 GPa and modulus of about 250 GPa. The modulus of high-modulus carbon fiber is more than 300 GPa. If the strength is greater than 4.0 GPa, carbon fiber is also called ultra-high-strength carbon fiber. Carbon fiber with modulus greater than 450 GPa is also called ultra-high-modulus carbon fiber. High-strength and high-elongation carbon fiber not only has high strength but also has good plasticity, and its elongation is more than 2%. Table 1.2 lists the types and properties of different carbon fibers.

1.2.1.2 SiC Fiber

SiC fibers are made from silicone compounds by the spinning, carbonization, or vapor deposition β-inorganic fibers with SiC structure. In terms of morphology, it can be divided into whisker and continuous fiber. The diameter of a silicon carbide whisker is generally approximately 0.1–2 μm. The length is approximately 20–300 μm. Continuous SiC fiber is formed by coating SiC on core fibers, such as tungsten wire or carbon fiber. Table 1.3 lists the types and properties of different SiC fibers.

1.2.1.3 Al₂O₃ Fiber

Alumina fiber is polycrystalline Al_2O_3 fiber, which is used as reinforcement. It has excellent mechanical strength and heat resistance. Its strength still decreases little until 1370°C. The composition and properties of various alumina fibers are listed in Table 1.4. The strength and other properties of alumina fiber mainly depend on its microstructure, such as pores, defects, and grain size, which have a significant impact on the properties of the fiber, while the microstructure of the fiber mainly depends on the preparation method and process of the fiber.

1.2.2 Fabric Architecture

According to the different braided structures of fiber preforms, CMCs can be divided into one-, two-, two-and-a-half-, and three-dimensional (1D, 2D, 2.5D, and 3D, respectively). Figure 1.1a shows a perspective view of plain-woven laminated 2D preform. The X and Y directions in the figure

TABLE 1.2   Type and Properties of Carbon Fibers

Type	Number of Fibers Per Bundle	Density (g/cm³)	Tensile Strength (MPa)	Tensile Modulus (GPa)	Tensile Elongation (%)	1 km Mass (g/km)
T300	1000	1.76	3530	230	1.5	66
	3000					198
	6000					396
	12000					800
T300J	3000	1.82	4410	230	1.9	198
	6000					396
	12000					800
T400H	3000	1.80	4410	250	1.8	198
	6000					396
T700S	12000	1.82	4800	230	2.1	800
T800H	6000	1.81	5900	294	1.9	223
	12000					445
T1000	12000	1.82	7060	294	2.4	448
T1000G	12000	1.80	6370	294	2.1	485
M35J	6000	1.75	5000	343	1.6	225
	12000					450
M40J	6000	1.77	4400	377	1.2	225
	12000					450
M46J	6000	1.84	4200	436	1	223
	12000					445
M50J	6000	1.87	4020	475	0.8	215
M55J	6000	1.93	3630	540	0.7	212
M6OJ	3000	1.94	3820	588	0.7	100
	6000					200
M30	1000	1.7	3920	294	1.3	53
	3000					160
	6000					320
	12000					640
M40	1000	1.81	2740	392	0.7	61
	3000					182
	6000					364
	12000					728
M46	6000	1.88	2550	451	0.6	360
M50	1000	1.91	2450	490	0.5	60
	3000					180

TABLE 1.3   Types and Properties of SiC Fibers

Properties	Nicalon NL-200	Hi-Nicalon	Hi-Nicalon Type S	Tyranno LoxM	Tyranno SA	Sylramic
Diameter/ [μm]	14	14	12	11	10	10
Fiber bundle/ [number of fibers per bundle]	500	500	500	800	1600	800
Tensile strength/ [GPa]	3.0	2.8	2.6	3.3	2.8	3.2
Tensile modulus/ [GPa]	220	270	420	187	420	380
Tensile elongation/ [%]	1.4	1.0	0.6	1.8	0.7	0.7
Density/ [g/cm^3]	2.55	2.74	3.10	3.3	3.24	3.55

TABLE 1.4   Types and Properties of Al$_2$O$_3$ Fibers

Type	Diameter/[μm]	Density/[g/cm^3]	Tensile Strength/[MPa]	Tensile Modulus/[GPa]
Nextel 312	10~12	2.7~2.9	1750	157
Nextel 440	10~12	3.5	2100	189
Nextel 480	10~12	3.05	2275	224
FP	20	3.9	1373	382
PRI166	20	4.2	2100~2450	385
TYC	250	3.99	2400	460

are the laying direction of the fiber bundle, and the Z direction is the laying direction of the fiber cloth. Figure 1.1b shows an X–Z plane sectional view of the 2D C/SiC composite.

Figure 1.2 shows the structural diagram of 3D preform CMC material. The 3D braided preform is shown in Figure 1.2a, which belongs to a 3D four-directional braided structure. The 3D C/SiC after preliminary matrix densification is shown in Figure 1.2b. The 3D braided preform can

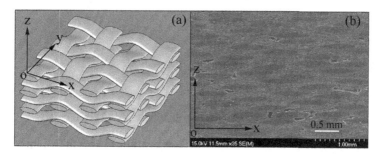

FIGURE 1.1 (a) 2D plain-woven fiber preform and (b) the X–Z plane sectional view of the 2D C/SiC composite.

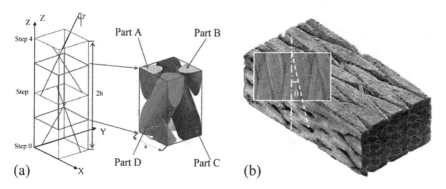

FIGURE 1.2 (a) 3D 4-directional fiber preform and (b) 3D C/SiC composite.

significantly improve the strength and stiffness of composites and has excellent mechanical properties and ablation resistance.

1.2.3 Interface

Generally, there is an obvious interface layer between the fiber of the preform and the SiC matrix as the transition layer. The interface layer is an important part of CMCs. The interface layer connects the fiber and ceramic matrix, which is the transition layer between the fiber and matrix material. Generally, pyrolytic carbon (PyC) and boron nitride (BN) can be used as the interface materials of CMCs. Figure 1.3 shows the 2D C/SiC composite with the PyC interphase.

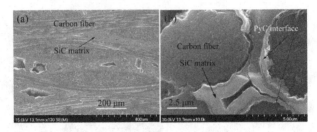

FIGURE 1.3 (a) 2D C/SiC composite and (b) the carbon fiber with PyC interphase.

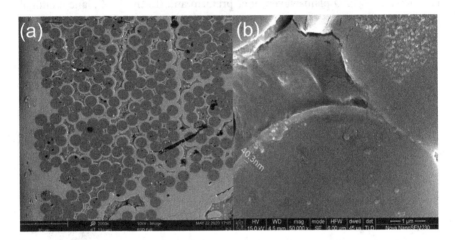

FIGURE 1.4 (a) CVI-C/SiC minicomposite and (b) PyC interphase.

1.2.4 Matrix

1.2.4.1 Chemical Vapor Infiltration

The chemical vapor infiltration (CVI) process includes two processes: interface layer deposition and matrix densification on the surface of 2D or 3D fiber preform. During deposition, the gas precursor is introduced into the furnace, and the gas reactant penetrates into the preform. At first, the interface is formed on the fiber surface, and then the SiC matrix is formed. Finally, the densification is realized. The obtained CMC usually contains 10–20% porosity. Figure 1.4 shows the CVI-C/SiC minicomposite and the carbon fiber with the PyC interphase.

Figure 1.5 shows the stress-displacement curves of fiber filament (FF) and nonwoven cloth (NC) CVI T-700™ C/SiC mini-composite samples with different interphases and reinforcements [3]. From the curves, all C/SiC minicomposites exhibit nonlinear characteristics and brittle

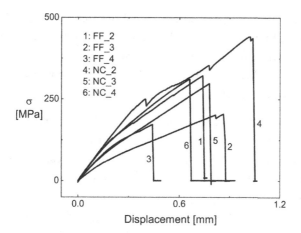

FIGURE 1.5   Load-displacement curves of C/SiC mini-composites with different interphase.

fracture behavior. For the C/SiC composite without the interphase, the tensile behavior of the composite exhibits linear fracture behavior, as the cracking in the matrix would penetrate through the fiber.

- For FF_2 C/SiC with a 6-h single-layer PyC interphase, the tensile curve of the minicomposite exhibits linear elastic till the proportional limit stress (PLS) of approximately σ_{PLS} = 95 MPa, and the tensile curve appears nonlinear due to the matrix cracking and interface debonding until the applied stress of approximately σ = 176 MPa, and then the tensile curve is linear elastic again until tensile fracture at the strength of approximately σ_{UTS} = 321.9 MPa.

- For FF_3 C/SiC with an 18-h single-layer PyC interphase, the tensile curve of the minicomposite exhibits linear elastic until the PLS of approximately σ_{PLS} = 40 MPa, and the tensile curve appears nonlinear due to matrix cracking and interface debonding until the applied stress of approximately σ = 120 MPa, and then the tensile curve is linear elastic again until tensile fracture at the strength of approximately σ_{UTS} = 204.6 MPa. Before tensile fracture, the tensile curve shows an obvious zigzag due to fibers' fracture.

- For FF_4 C/SiC with 4-layer PyC-SiC interphase, the tensile curve of the minicomposite exhibits linear elastic until the PLS of approximately σ_{PLS} = 89 MPa, and the tensile curve appears nonlinear due to matrix cracking and interface debonding until the applied stress of approximately $\sigma = 138$ MPa, and then the tensile curve is linear elastic again until tensile fracture at the strength of approximately $\sigma_{UTS} = 172.2$ MPa. Under tensile loading, the tensile curve does not show zigzag behavior.

- For NC_2 C/SiC with 6 h single layer PyC interphase, the tensile curve of the minicomposite exhibits linear elastic until the PLS of approximately $\sigma_{PLS} = 196$ MPa, and with increasing load, the zigzag behavior occurs at the applied stress $\sigma = 249$, 353 and 441 MPa, respectively, due to matrix cracking and fiber fracture, and the composite tensile fracture occurs at the strength of approximately $\sigma_{UTS} = 441.5$ MPa.

- For NC_3 C/SiC with an 18-h single-layer PyC interphase, the tensile curve of the minicomposite exhibits linear elastic until the PLS of approximately $\sigma_{PLS} = 72$ MPa, and the tensile curve appears nonlinear due to matrix cracking and interface debonding until the applied stress of approximately $\sigma = 141$ MPa, and then the tensile curve is linear elastic again until tensile fracture at the strength of approximately $\sigma_{UTS} = 298.4$ MPa. There is no zigzag appearance under tensile loading.

- For NC_4 C/SiC with a 4-layer PyC-SiC interphase, the tensile curve of the minicomposite exhibits linear elastic until the PLS of approximately $\sigma_{PLS} = 94$ MPa, and with increasing load, the zigzag behavior occurs at the stress of $\sigma = 130$ MPa mainly due to the matrix cracking, and with the continually increasing load, the tensile curve shows nonlinear behavior until tensile fracture at the strength of approximately $\sigma_{UTS} = 311.9$ MPa.

With respect to the damage mechanism, three or four stages may occur under tensile load:

- Stage I, an elastic response coupled with a partial reopening of thermal microcracking

- Stage II, multiple matrix micro-cracking perpendicular to the applied loading

- Stage III, crack opening and related fiber/matrix and mostly bundle/matrix and inter-bundle debonding

- Stage IV, progressive transfer of load to the fiber and gradual fiber failure until composite failure/fracture

In addition, it is particularly interesting for the sample NC_2 that the curve of stress–displacement exhibits a zigzag behavior, which means that part of the carbon FFs were broken first and the rest of filaments continue to bear the increasing load. In fact, this sample is made up of two carbon fiber bundles, which have stuck together during the process of CVI SiC densification, as shown in Figure 1.6. Two bundles carbon fibers are marked by blue rectangle box and red ellipse box. From the fracture section, a fracture step between the bundles can be clearly seen, which means that two bundles did not break synchronously under load. This phenomenon can also be found in the sample FF_2, which is just one bundle carbon fiber as reinforcement. From the curve of FF_2, some discontinuous changes can be noticed as well, which explains that all the filaments in the whole bundle cannot simultaneously carry the load. It is worth mentioning that the bundle integrity plays another important role in the mechanical properties. Lacking bundle integrity, sample FF_2 cannot take the load

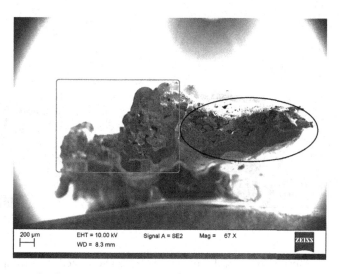

FIGURE 1.6   The fracture section of sample NC_2 C/SiC minicomposite.

as a union and the fracture of minicomposite exhibits progressive damage during the tensile test, which implies that partial fiber filaments fracture first and then the rest of the filaments disrupt under further loading, as shown in Figure 1.7.

1.2.4.2 Polymer Infiltration and Pyrolysis

The SiC fiber is woven into the preform configuration through 2D and 3D fabrics with the help of the mold, and then the preform is prepared with the interface layer. Then the preform is placed in the solution of

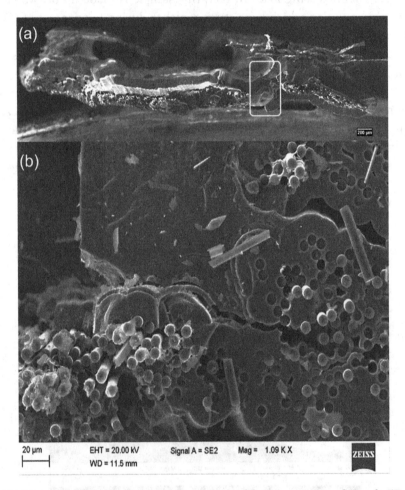

FIGURE 1.7 Morphologies and topographies of fracture section of sample FF_2 C/SiC minicomposite (a) overall morphology of fracture surface and (b) local morphology of amplification.

the precursor for impregnation. Finally, high-temperature pyrolysis (the pyrolysis temperature is generally 700–1600°C) converts the polymer into ceramic materials. The density of the preform is correspondingly increased from 1 g/cm^3 to 2–3 g/cm^3. In the process of cyclic impregnation and pyrolysis, the matrix shrinks and forms microcracks, usually 5–10 cycles are required for densifying the material.

Figure 1.8 shows the 3D microstructure of polymer infiltration and pyrolysis (PIP)–SiC/SiC minicomposite observed under X-ray [4]. The SiC fibers are not uniformly distributed inside the matrix, some fibers are not

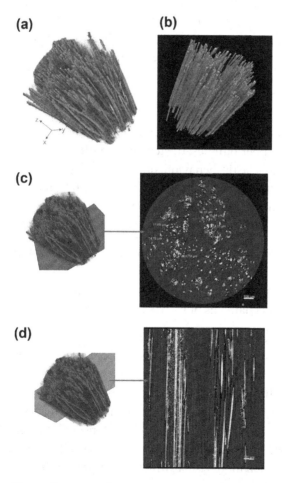

FIGURE 1.8 Three-dimensional microstructure of PIP-C/SiC composite observed under X-ray (a) grayscale image, (b) rendering image, (c) X–Y plane, and (d) X–Z plane.

straight, and there is a bending phenomenon in the reinforcing fibers. From the observations along the X–Y and X–Z plane, there is matrix-rich region inside the minicomposite. It should be noted that the yellow region is identified as the reinforcing fibers and the blue region is identified as the matrix. The fiber distribution changes along the length of the specimen. There exists a matrix-rich region between the reinforcing fibers. As the failure strain of the SiC matrix is much less than that of the SiC fiber, matrix cracking occurs at low applied stress, and when the microcracking caused by thermal residual stress propagates to the matrix-rich region, the microcracking can propagate and transfer into the long matrix cracking.

Figure 1.9 shows the load–displacement curves of original BX™ SiC fiber bundles, processed and uncoated BX™ SiC fiber bundles, PIP BX™ SiC/SiC minicomposite without the PyC interphase, BX™ SiC fiber bundles with the single-layer PyC interphase, and PIP BX™ SiC/SiC minicomposite with the single-layer PyC interphase.

- For the original BX™ SiC fiber bundles and processed and uncoated BX™ SiC fiber bundles, the tensile load–displacement curves exhibited linear-elastic behavior, due to the absence of micro-damage mechanisms of matrix fragmentation and interface debonding, and approaching tensile failure, the tensile curves show nonlinear behavior mainly due to gradual fibers' fracture in the bundles.

- For the PIP BX™ SiC/SiC minicomposite without interphase, the tensile load–displacement curves show a broad dispersion and nonlinear behavior at low applied load, which is mainly attributed to the micro-damage mechanisms of matrix fragmentation during fabrication.

- For the BX™ SiC fiber bundles with a single-layer PyC interphase, the tensile load-displacement curves display a nonlinear elastic behavior, and approaching tensile fracture, the tensile curves show a marked zigzag behavior, which is mainly attributed to fibers' gradual fracture. Compared with the original and with processed and uncoated BX™ SiC fiber bundles, the zigzag behavior near the tensile fracture for BX™ SiC fiber bundle with the single-layer PyC interphase at high stress is much more visible, which may be due to the failure of clusters.

- For the PIP BX™ SiC/SiC minicomposite with the PyC interphase, the tensile load–displacement curves show a clear nonlinear behavior.

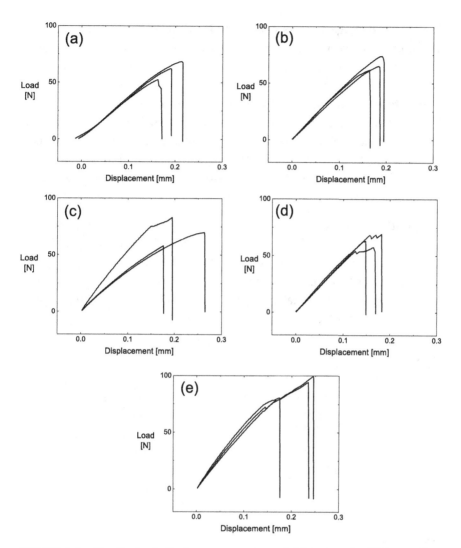

FIGURE 1.9 Tensile load-displacement curves of (a) original BX™ SiC fiber bundles, (b) processed and uncoated BX™ SiC fiber bundles, (c) PIP BX™ SiC/SiC minicomposite without the interphase, (d) BX™ SiC fiber bundles with the single-layer PyC interphase, and (e) PIP BX™ SiC/SiC minicomposite with the single-layer PyC interphase.

The tensile damage process can be divided into three regions, and the relationship between the tensile curves and microstructure damage is established:

- Before tensile loading, the specimen surface exhibits the appearance of a marked matrix fragmentation, which was caused during the fabrication, as shown in Figure 1.10a. At low applied stress (i.e., $\sigma < 100$ MPa), these matrix fragmentations caused during processing propagated, and the tensile load–displacement curves exhibited linear-elastic behavior at a low-stress range.

- With increasing applied stress, microcracking caused during processing propagates to the matrix-rich region and the fiber/matrix interface, and the long matrix crack appears inside the minicomposite and the tensile loading–displacement curves show a nonlinear behavior at the load of approximately 70–80 N (corresponding to approximately $\sigma = 170 - 180$ MPa), as shown in Figures 1.9e and 1.10b.

(a)

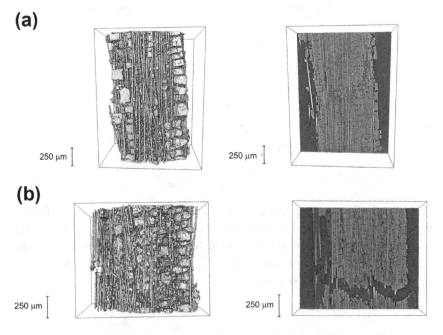

(b)

FIGURE 1.10   Tensile damage observation of PIP BX™ SiC/SiC with PyC interphase under X-ray in X–Z plane (a) original specimen and (b) tensile fracture specimen with matrix fragmentation and fiber pullout.

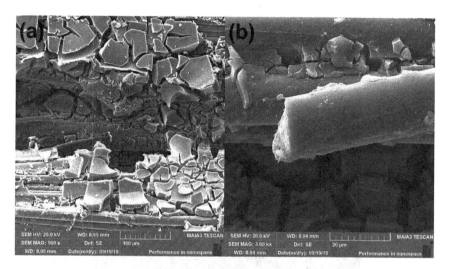

FIGURE 1.11   SEM schematic of PIP BX™ SiC/SiC minicomposite with the PyC interphase (a) matrix fragmentation and (b) fiber pullout.

- The minicomposite tensile fracture occurred at the long matrix cracking region with fiber fracture and pullout, as shown in Figure 1.10b. At the fracture surface, multiple matrix fragmentation and fiber pullout can be observed, as shown in Figure 1.11.

1.2.4.3 Melt Infiltration (MI)

To improve a composite's toughness, a thin BN interphase produced from boron trichloride (BCl_3) and ammonia (NH_3) was deposited first on the surfaces of SiC fibers by means of the CVI process [5]. The thickness of the interphase was measured under scanning electron microscopy (SEM) and was approximately 300–500 nm, as shown in Figure 1.12. The coated SiC fibers were then wetted by phenolic resin to transform into SiC pre-pregs. After cutting the prepregs and laminating them into a steel mold, a hot-pressing (HP) procedure was performed to obtain green bodies via thermal curing of the phenolic resin. The green bodies were subsequently pyrolyzed to convert the phenolic resin into a carbon source, which was used for the following silicon infiltration process to form the SiC matrix by the reaction of carbon with silicon. Furthermore, the surfaces of prepared SiC/SiC composite panels were slightly ground by diamond polishing plates to remove the excess silicon after the siliconization process.

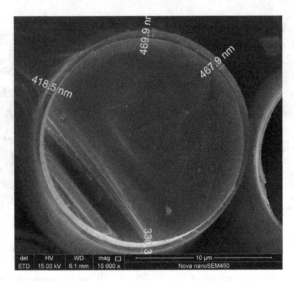

FIGURE 1.12   BN interphase on SiC fiber.

FIGURE 1.13   Surface of MI-SiC/SiC composite observed under SEM.

Figure 1.13 shows the surface of MI-SiC/SiC composite, and it can be found that there are no cracks on the surface of the composite.

Figure 1.14 shows the experimental tensile stress-strain curves for unidirectional and cross-ply Cansas-3203™ MI-SiC/SiC composites [5]. The tensile stress-strain curves in unidirectional and cross-ply MT-SiC/SiC composites exhibit the same trend but have large variations in applied stress and strain values, mainly due to the damage mechanisms of transverse cracking in 90° plies and low ECFL ($\psi = 0.5$) for cross-ply laminates.

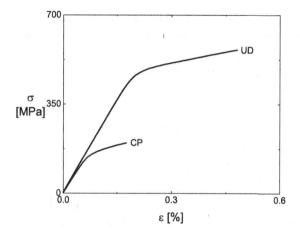

FIGURE 1.14   Experimental tensile stress-strain curves for unidirectional and cross-ply MI-SiC/SiC composites at room temperature.

The tensile stress-strain curves of these composites showed obvious non-linearity, revealing three stages:

- **Stage I.** The composite's tensile stress-strain curves exhibited a linear-elastic pattern, and micro-matrix or transverse cracking in the 90° plies may occur. However, these microcracks in matrix or transverse cracks in 90° plies did not affect the tensile linear-elastic behavior of unidirectional or cross-ply MI-SiC/SiC composites. For the unidirectional MI-SiC/SiC composite, the stress range for the Stage I was $\sigma \in$ [0, 280 MPa], with the corresponding strain range $\varepsilon_c \in$ [0, 0.11%]; however, for the cross-ply MI-SiC/SiC composite, the stress range for the Stage I was $\sigma \in$ [0, 40 MPa], with the corresponding strain range $\varepsilon_c \in$ [0, 0.017%].

- **Stage II.** The composite's tensile stress-strain curves include the linear and nonlinear segments. With the propagating of existing matrix cracks and more new matrix cracks occurring, the tensile stress-strain curves gradually transfer from linear to nonlinear. For the unidirectional MI-SiC/SiC composite, the stress range for the damage Stage II was $\sigma \in$ [280 MPa, 480 MPa], with the corresponding strain rage $\varepsilon_c \in$ [0.11%, 0.22%]; however, for the cross-ply SiC/SiC composite, the stress range for the damage Stage II was $\sigma \in$ [40 MPa, 160 MPa], with the corresponding strain range $\varepsilon_c \in$ [0.017%, 0.089%].

- **Stage III**. The composite's tensile stress-strain curves exhibited a secondary quasi-linear elastic pattern accompanied by saturated matrix cracking and gradual fibers fracture. Due to the high tensile strength of the SiC fibers, gradual fiber fracture mainly occurs at Stage III. For the unidirectional SiC/SiC composite, the stress range for damage in Stage III was $\sigma \in$ [480 MPa, 562 MPa], with the corresponding strain range $\varepsilon_c \in$ [0.22%, 0.48%]; however, for the cross-ply MI-SiC/SiC composite, the stress range for damage in Stage II was $\sigma \in$ [160 MPa, 197.8 MPa], with the corresponding strain rage $\varepsilon_c \in$ [0.089%, 0.176%].

Figure 1.15 shows the cyclic loading/unloading tensile curves of unidirectional and cross-ply MI-SiC/SiC composites at ambient. Upon cyclic loading/unloading, tensile stress-strain curves showed apparent hysteresis loops. With tensile peak stress, hysteresis loops' areas increased; while approaching tensile peak stress, non-closure hysteresis loops appeared, which may be due to matrix fragmentation progressing with higher stress levels. Under cyclic compliance tensile, *in situ* acoustic emission (AE) was utilized for monitoring internal damage evolution in composites.

- For the unidirectional MI-SiC/SiC composite, in damage in Stage I, there was nearly no AE signal. When the tensile peak stress increased to $\sigma_{max} = 280$ MPa, a low AE signal, with a peak AE energy $\Delta_{max} = 223$ mV·μs was found, indicating the initiation of matrix microcracking and the beginning of damage in Stage II. When the tensile peak stress rose to approximately $\sigma_{max} = 360$ and 400 MPa, the peak AE energy

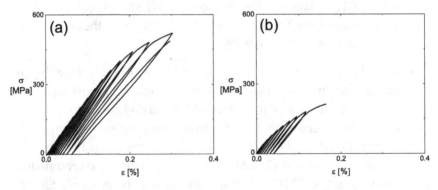

FIGURE 1.15 Cyclic loading/unloading tensile curves of (a) unidirectional and (b) cross-ply MI-SiC/SiC composites at ambient.

was increased to Δ_{max} = 9132 and 39380 mV·μs, respectively. With the tensile peak stress increasing to approximately σ_{max} = 440 MPa, which approached the PLS, the peak AE energy greatly increased to Δ_{max} = 70217 mV·μs. However, after the peak stress rose from 480 to 520 MPa, the AE energy remained almost constant, indicating the saturation of matrix fragmentation and gradual fiber facture, indicating the beginning of damage in Stage III.

- For the cross-ply MI-SiC/SiC composite, no AE signal was observed at low applied stress, that is, σ_{max} < 20 MPa, indicating there was no damage inside the composite at Stage I. However, at σ_{max} = 40 MPa, a low AE signal was observed with a peak AE energy of Δ_{max} = 673 mV·μs, indicating the initiation of matrix fragmentation and the beginning of damage in Stage II. When the peak stress was increased to σ_{max} = 140 MPa, the peak AE energy slowly increased to Δ_{max} = 7808 mV·μs. With peak stress increased to σ_{max} = 160 MPa, the peak AE energy was increased to Δ_{max} = 10572 mV·μs. With σ_{max} increased to 180 MPa, AE signal was greatly increased to Δ_{max} = 26233 mV·μs, indicating the saturation of matrix fragmentation inside the composite at Stage II. Approaching tensile fracture, a large number of fibers were broken with pullout from the matrix; at this time, AE energy suddenly rose to Δ_{max} = 51554 mV·μs with the damage in Stage III.

Compared with cross-ply MI-SiC/SiC composite, the tensile peak stress corresponding to AE signal occurrence was much higher for the unidirectional MI-SiC/SiC composite, that is, 20–40 MPa for cross-ply composite versus 240–280 MPa for unidirectional composite. Approaching composite's proportional limit stress, the AE signals of both unidirectional and cross-ply composites were increased rapidly, indicating a rapid increase in matrix fragmentation. Upon reloading, there was no AE signal at the beginning of reloading, indicating there was no propagation of matrix cracking or new matrix cracks, due to the Kaiser effect; however, with the gradually increasing tensile stress AE signals, high-AE energy signals appeared only with tensile stress exceeding the previous peak level.

Figure 1.16 depicts the SEM observation of fracture surface for unidirectional MI-SiC/SiC composite. The fracture surface had a step shape with matrix cracks. With matrix cracks propagating into the fiber, the cracks deflected along the interphase, leading to the interface debonding.

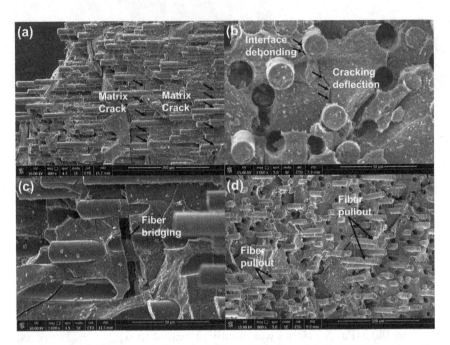

FIGURE 1.16 Fracture surface for the unidirectional MI-SiC/SiC composite observed under SEM. (a) Matrix cracks, (b) interface debonding and crack deflection, (c) fiber bridging, and (d) fiber pullout.

Due to the deflection of cracks, the propagation path was extended, which prevented the fibers from being directly fractured by the cracks and delayed the fracture process of the matrix. Because of the weak bonding strength between the fibers and the interface layer, the fibers debonded from the interface layer under the tensile stress and could bridge the microcracks in the matrix, reducing the crack opening.

Figure 1.17 depicts the SEM observation of the fracture surface for cross-ply MI-SiC/SiC composite. Under tensile loading, matrix cracks were formed in the 0° plies. The left matrix crack in Figure 1.17a is not clear compared with the right matrix crack, due to the closure of matrix cracks. The transverse cracks in the 90° plies connected the matrix cracks in the 0° plies to form the major cracks. These major cracks increased the load on the fibers, leading to the high fracture probability for the fibers in the 0° plies. However, the pullout length of the cross-ply MI-SiC/SiC composite was much shorter in comparison with that of the unidirectional MI-SiC/SiC composite, mainly due to the transverse matrix cracking in the 90° plies. Transverse cracks in the 90° plies lead to a high-stress

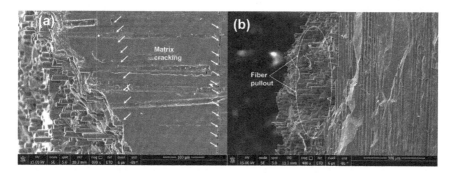

FIGURE 1.17   Fracture surface for the cross-ply MI-SiC/SiC composite observed under SEM. (a) Matrix cracks and (b) fiber pullout.

concentration in the 0° plies, which increases the fiber fracture probability and decreases the fiber pullout length.

1.2.4.4 HP

For C/SiC composite fabricated using the HP method, the nano-SiC powder and sintering additives were ball-milled for 4 h using SiC balls. After drying, the powders were dispersed in xylene with polycarbonsilane (PCS) to form the slurry. Carbon fiber tows were infiltrated by the slurry and wound to form aligned unidirectional composite sheets. After drying, the sheets were cut and pyrolyzed in argon. Then the sheets were stacked in a graphite die and sintered by HP [6, 7].

The tensile stress-strain curve of unidirectional HP-C/SiC is shown in Figure 1.18. The composite behaves as a typical damageable material, exhibiting an extended nonlinear stress-strain domain up to rupture. As the large mismatch of the axial thermal expansion coefficients between the carbon fibers and SiC matrix ($-0.38 \times 10^{-6}/°C$ vs $2.8 \times 10^{-6}/°C$) and the resultant large residual tensile stresses in SiC matrix, there are unavoidable microcracks existed within the SiC matrix when the composites are cooled from the high process temperature to room temperature. However, an initial linear region (below ~50 MPa) still exists, and the initial elastic modulus of the composite is about 170 GPa. These processing-induced microcracks propagated when the tensile loading increased in conjunction with new microcracks initiated in the matrix. These microcracks joined together to form macrocracks, and some microcracks deflected along the fiber/matrix interface. Multiple matrix cracking and fiber/matrix interface debonding lead to the macroscopic nonlinearity of composites.

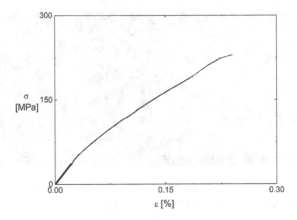

FIGURE 1.18   Tensile stress-strain curve of unidirectional HP-C/SiC composite.

The matrix cracks approach saturation at approximately 180–200 MPa. After the matrix cracks reach saturation, the fibers carry the extra loads during the continued loading. The specimen finally failed at stress about 230 MPa, with a corresponding failure strain of 0.24%. The optical micrograph of the multiple matrix cracking on the edge surface and the fiber pullout at the fracture surface for the failed tensile specimen is shown in Figure 1.19. The average saturation matrix crack spacing is about 106 μm. The material exhibited extensive fiber pullout during tensile failure (Figure 1.2b), with distributed fiber failure.

The tensile stress-strain curve of cross-ply HP-C/SiC composite is shown in Figure 1.20. The composite behaves as a typical damageable

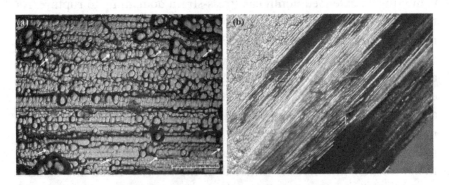

FIGURE 1.19   The failed tensile specimen observed under optical microscope (a) the matrix multiple cracking on the edge surface and (b) the fiber pullout at the fracture surface.

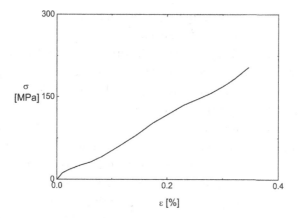

FIGURE 1.20   Tensile stress-strain curve of cross-ply HP-C/SiC composite.

material, exhibiting an extended nonlinear stress–strain domain up to failure. There are unavoidable opening microcracks existed in SiC matrix at different plies as the high thermal residual tensile stress in the SiC matrix lies in the region of statistical distribution of strength, that is, a two-parameter Weibull distribution. By increasing the applied stress, these microcracks would propagate, and new matrix crack would appear in low-matrix-strength region. During the process of multiple matrix micro-cracks under low stress levels, thermal stresses have partially relaxed. The initial linear elastic region (below ~20 MPa) still exists, and the initial elastic modulus of the composite is ~88 GPa. These processing-induced micro-cracks propagated with the increase of tensile stress, in conjunction with new transverse cracks in the 90° plies. With the increase of applied stress, transverse crack spacing decreases and approaches saturation. When the applied stress increases to approximately 30 MPa, matrix cracking and fiber/matrix interface debonding occur in the 0° plies, and the tensile stress-strain curve behaves linear again with the tangent elastic modulus of approximately 64 GPa. Upon continually increase of the applied stress to about 140 MPa, the tensile stress-strain curve deviates from linear again due to saturation of matrix cracking in 0° plies. The tensile stress-strain curve behaves linearly with a tangent elastic modulus of approximately 76 GPa. After the matrix cracks approach saturation, the fibers begin to fracture. The ultimate tensile strength of cross-ply HP-C/SiC composite is about 204 MPa, with corresponding failure strain of approximately 0.35%.

With the optical micrograph of transverse cracking, matrix cracking, and major cracking on the edge surface, the fiber pullout on the fracture surface of tensile failure specimen is shown in Figure 1.21. The transverse cracks in the 90° plies on the edge of specimen are shown in Figure 1.21a. The transverse cracks propagated through the thickness of 90° plies, and the crack spacing is approximately 218 μm. The matrix cracks in the 0° plies on the edge of the specimen are shown in Figure 1.21b. The matrix cracks propagated through the thickness of 0° plies, and the crack spacing is about 129 μm. Comparing Figure 1.21a with Figure 1.21b, the transverse crack spacing is much larger than matrix crack spacing. The major cracking, that is, transverse cracking connects with matrix cracking, is shown in Figure 1.21c. The major cracks propagate through the 0° and 90° plies. Comparing Figure 1.21b with Figure 1.21c, it is proposed that matrix microcracks do not necessarily extend from existing transverse cracks. Some matrix cracks formed in the matrix-rich regions and then

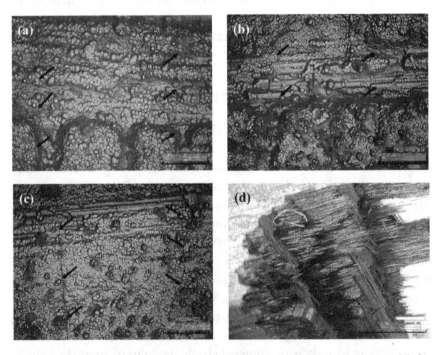

FIGURE 1.21 Tensile failure specimen observed under optical microscope (a) the transverse cracks in the 90° plies, (b) the matrix cracks in the 0° plies, (c) the major cracks propagating through the 90° and 0° plies, and (d) the fiber pullout at the fracture surface.

multiplied throughout the matrix with an increase of applied stress. The fracture surface of failure specimen is shown in Figure 1.21d. The fracture surface of the 0° plies appears like a sawtooth, in which the bridging fiber pulled out from the matrix. However, fibers in the 90° plies did not fracture as these fibers are perpendicular to the loading direction. After the saturation of transverse cracks and matrix cracks, the tensile load is mainly carried by intact fibers of 0° plies. The fibers in 90° plies are perpendicular to the load, which behaves with weak load-carrying ability.

1.3 HIGH-TEMPERATURE MECHANICAL HYSTERESIS BEHAVIOR IN DIFFERENT CMCS

1.3.1 Unidirectional C/SiC

The tension-tension fatigue behavior of unidirectional HP-C/SiC composite was investigated at 800°C in an air atmosphere [8]. The fatigue experiments were in a sinusoidal waveform and a loading frequency of f = 10 Hz. The fatigue load ratio was R = 0.1, and the maximum number of applied cycles was defined to be 10^6 cycles.

Under σ_{max} = 250 MPa, the experimental fatigue hysteresis loops corresponding to the 1st, 1000th, 5000th, 10000th, 15000th 20000th, and 24000th cycles are shown in Figure 1.22a. The theoretical hysteresis dissipated energy versus the interface shear stress curve is shown in Figure 1.22b, in which the hysteresis dissipated energy increases with a decrease in the interface shear stress to the peak value ΔW = 90 kPa (the corresponding interface shear stress is τ_i = 4.8 MPa) and then decreases with

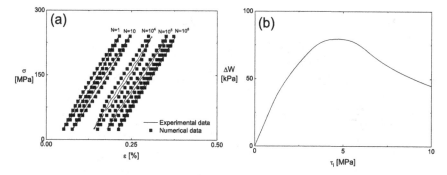

FIGURE 1.22 (a) Experimental and predicted fatigue hysteresis loops and (b) theoretical fatigue hysteresis dissipated energy as a function of interface shear stress of unidirectional C/SiC composite under σ_{max} = 250 MPa at 800°C in an air atmosphere.

decrease of the interface shear stress to $\Delta W = 0$ kPa (the corresponding interface shear stress is $\tau_i = 0$ MPa). Experimental hysteresis dissipated energy of the 1st, 100th, 1000th, 5000th, 10000th, 15000th, 20000th, and 24000th cycles are 62, 50, 24, 16, 12, 8, 7.8, and 7.2 kPa, respectively. Under $\sigma_{max} = 250$ MPa, the hysteresis dissipated energy of the first cycle lies in the right part of the hysteresis dissipated energy versus the interface shear stress curve. The hysteresis loop of the first cycle corresponds to interface slip Case 2; that is, the interface partially debonds, and the fiber partially slides relative to the matrix in the interface debonding region upon unloading/reloading. When the interface completely debonds, the interface shear stress degrades rapidly due to the interface radial thermal residual tensile stress. The hysteresis loop of the 100th cycle corresponds to interface slip Case 4; that is, the interface completely debonds, and the fiber completely slides relative to the matrix in the interface debonding region upon unloading/reloading.

1.3.2 Cross-Ply C/SiC and SiC/MAS-L Composites

The tension–tension fatigue behavior of cross-ply C/SiC composite was investigated at 800°C in air atmosphere. The fatigue experiments were in a sinusoidal waveform and a loading frequency of $f = 10$ Hz. The fatigue load ratio was $R = 0.1$, and the maximum number of applied cycles was defined to be 10^6 cycles.

Under $\sigma_{max} = 105$ MPa, the fatigue hysteresis loops of the 4th, 10th, 100th, 500th, 1000th, and 6000th cycles are given in Figure 1.23a. The theoretical

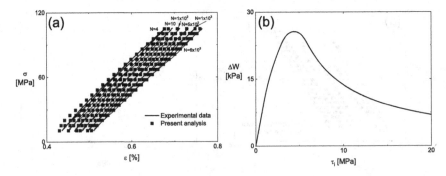

FIGURE 1.23 (a) Experimental and predicted fatigue hysteresis loops and (b) theoretical fatigue hysteresis dissipated energy as a function of interface shear stress in the 0° ply of cross-ply C/SiC composite under $\sigma_{max} = 105$ MPa at 800°C in an air atmosphere.

hysteresis dissipated energy versus the interface shear stress curve is shown in Figure 1.23b, in which the hysteresis dissipated energy increases with a decrease in interface shear stress to the peak value ΔW = 25.6 kPa (the corresponding interface shear stress is τ_i = 4.4 MPa) and then decreases with decrease of the interface shear stress to ΔW = 0 kPa (the corresponding interface shear stress is τ_i = zero MPa). The experimental hysteresis dissipated energy of the 1st, 2nd, 3rd, 4th, 10th, 100th, 500th, 1000th, 3000th, 6000th and 6600th cycles are 24.3, 20, 13, 12, 9.7, 8.6, 7.1, 6.1, 5.4, 5.2 and 5.1 kPa, respectively. Under σ_{max} = 105 MPa, the hysteresis dissipated energy of the first cycle lies in the right part of the hysteresis dissipated energy versus the interface shear stress curve. The hysteresis loop of the first cycle corresponds to interface slip Case 2; that is, the interface partially debonds, and the fiber partially slides relative to the matrix in the interface debonding region upon unloading/reloading. The hysteresis loop of the 100th cycle corresponds to interface slip Case 4; that is, the interface completely debonds and the fiber completely slides relative to the matrix in the interface debonding region upon unloading/reloading.

The tension-tension fatigue behavior of cross-ply SiC/MAS-L composite at 800°C and 1000°C in an inert atmosphere was investigated. The fatigue loading frequency was f = 1 Hz. The fatigue peak and valley stresses were 110 and 0 MPa, respectively. The hysteresis loops corresponding to each cycle were recorded and analyzed. Experimental fatigue hysteresis dissipated energy versus the number of applied cycles curves is shown in Figure 1.24a. The theoretical hysteresis dissipated energy versus the

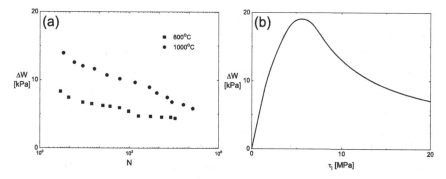

FIGURE 1.24 (a) Experimental fatigue hysteresis dissipated energy versus the number of applied cycles and (b) the fatigue hysteresis dissipated energy versus interface shear stress in the 0° ply of cross-ply SiC/MAS–L composite under σ_{max} = 110 MPa at T = 800°C and 1000°C in an inert atmosphere.

interface shear stress curve is shown in Figure 1.24b, in which the hysteresis dissipated energy increases with decrease of interface shear stress to the peak value ΔW = 19 kPa (the corresponding interface shear stress is τ_i = 5.6 MPa) and then decreases with decrease of interface shear stress to ΔW = 0 kPa (the corresponding interface shear stress is τ_i = 0 MPa). At 800°C in an inert atmosphere, the experimental hysteresis dissipated energy degrades from ΔW = 8.3 kPa at the 5th cycle to ΔW = 4.3 kPa at the 34162nd cycle, and at 1000°C in an inert atmosphere, the experimental hysteresis dissipated energy degrades from ΔW = 14 kPa at the 6th cycle to ΔW = 5.8 kPa at the 133925th cycle. As the radial thermal expansive coefficient of MAS-L matrix is lower than that of SiC fiber, the thermal residual tensile stress exists in the fiber/matrix interface, which lowers the interface shear stress. When the composite was cooled down from high fabricated temperature to room temperature, the fiber/matrix interface debonds. The radial thermal residual tensile stress decreases with an increase in the test temperature, leading to an increase in the interface shear stress with an increase of the test temperature. Under fatigue loading at elevated temperature, the hysteresis dissipated energy decreases with increase of the number of applied cycles, corresponding to the left part of the hysteresis dissipated energy versus interface-shear stress curve. The hysteresis loops correspond to interface slip Case 4; that is, the interface completely debonds and fiber completely slips relative to matrix in the interface debonded region upon unloading/reloading.

1.3.3 2D Plain-Woven SiC/SiC

The tension-tension fatigue behavior of 2D SiC/SiC composite at 600, 800, and 1000°C in an inert atmosphere was investigated. The fatigue loading frequency was f = 1 Hz. The fatigue peak and valley stresses were 130 and 0 MPa, respectively. Experimental hysteresis loss energy versus the number of applied cycles curves is shown in Figure 1.25(a). At T = 600°C in an inert atmosphere, the hysteresis dissipated energy increases from ΔW = 5.2 kPa at the 15th cycle to ΔW = 9.4 kPa at the 333507th cycle; at T = 800°C in an inert atmosphere, the hysteresis dissipated energy increases from ΔW = 9 kPa at the 23rd cycle to ΔW = 15.3 kPa at the 97894th cycle, and at T = 1000°C in an inert atmosphere, the hysteresis dissipated energy increases from ΔW = 10 kPa at the 22nd cycle to ΔW = 21.8 kPa at the 117055th cycle. The theoretical hysteresis dissipated energy versus the interface–shear stress curve is shown in Figure 1.25b. As the radial

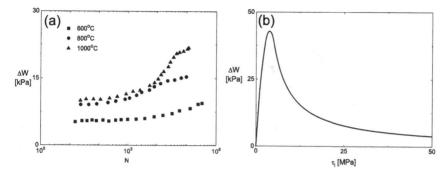

FIGURE 1.25 (a) Experimental fatigue hysteresis dissipated energy versus the number of applied cycles and (b) theoretical fatigue hysteresis dissipated energy versus interface shear stress in the longitudinal yarns of 2D SiC/SiC composite under σ_{max} = 130 MPa at T = 600, 800, and 1000°C in an inert atmosphere.

thermal expansive coefficient of SiC matrix is higher than that of SiC fiber, the radial thermal residual compressive stress exists in the fiber/matrix interface, which increases the interface shear stress. When the composite was cooled from high fabricated temperatures to room temperature, the radial thermal residual compressive stress and interface shear stress both decrease with an increase in the test temperature. Under fatigue loading at elevated temperatures, the hysteresis dissipated energy increases with an increase of the number of applied cycles, corresponding to the right part of the hysteresis dissipated energy versus the interface-shear stress curve. The hysteresis loops correspond to interface slip Case 2; that is, the interface partially debonds and the fiber partially slides relative to matrix in the interface debonding region upon unloading/reloading.

1.3.4 2.5D C/SiC

The tension-tension fatigue behavior of 2.5D C/SiC composite at 800°C in air atmosphere. The fatigue loading was in a sinusoidal waveform and a frequency of f = 10 Hz. The tensile fatigue stress ratio was R = 0.1, and the maximum number of applied cycles was defined to be 10^6 cycles.

Under σ_{max} = 140 MPa, the hysteresis loops at the 500th, 15000th, 20000th, and 22700th cycles are shown in Figure 1.26a. The specimen experienced 5281 cycles and then fatigue fractured. The theoretical hysteresis dissipated energy versus the interface shear–stress curve is shown in Figure 1.26b, in which the hysteresis dissipated energy increases with a decrease in the interface shear stress to the peak value ΔW = 21.7 kPa (the

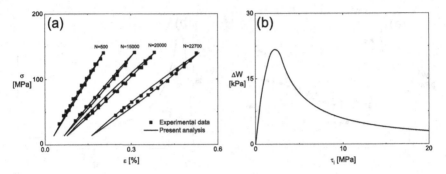

FIGURE 1.26 (a) Experimental and predicted fatigue hysteresis loops; and (b) Theoretical fatigue hysteresis dissipated energy as a function of interface shear stress in the longitudinal yarns of 2.5D C/SiC composite under $\sigma_{max} = 140$ MPa at 800°C in air atmosphere.

corresponding interface shear stress is $\tau_i = 2.25$ MPa) and then decreases with decrease of interface shear stress to $\Delta W = 0$ kPa (the corresponding interface shear stress is $\tau_i = 0$ MPa). Experimental hysteresis dissipated energy of the 500th, 15000th, 20000th, and 22700th cycles are 6.3, 7.2, 8.7, and 11.8 kPa, respectively. Under $\sigma_{max} = 140$ MPa, the hysteresis dissipated energy increases with an increase in the number of applied cycles. The hysteresis loops from the 500th cycle to 22700th cycle all correspond to the interface slip Case 2; that is, the interface partially debonds and the fiber partially slides relative to the matrix in the interface debonding region on unloading/reloading.

The tension-tension fatigue behavior of 2.5D C/SiC composite at 600°C in an inert atmosphere was investigated. The fatigue peak and valley stresses were 230 and 0 MPa, respectively. The fatigue loading frequency was $f = 1$ Hz. Experimental hysteresis loops corresponding to the 10th, 100000th, and 1000000th cycles are shown in Figure 1.27a. The theoretical hysteresis dissipated energy versus the interface shear stress curve is shown in Figure 1.27b. As the axial thermal expansion coefficient of carbon fiber is lower than that of SiC matrix, the axial thermal residual tensile stress exists in the fiber/matrix interface, leading to microcracking in SiC matrix upon being cooled from the high fabricated temperature to room temperature. The radial thermal expansion coefficient of carbon fiber is higher than that of SiC matrix, the radial thermal residual tensile stress exists in the fiber/matrix interface, leading to interface debonding upon

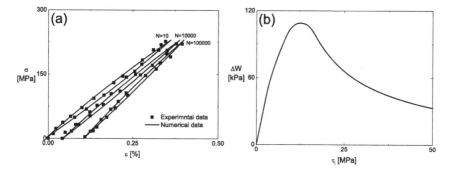

FIGURE 1.27 (a) Experimental fatigue hysteresis loops of different number of applied cycles and (b) theoretical fatigue hysteresis dissipated energy versus interface shear stress in the longitudinal yarns of 2.5D woven C/SiC composite under σ_{max} = 230 MPa at 600°C in an inert atmosphere.

being cooled from the high fabrication temperature to room temperature. Experimental hysteresis dissipated energy of the 10th, 10000th, and 100000th cycles are 33, 29, and 19 kPa, respectively. Under σ_{max} = 230 MPa, the hysteresis dissipated energy decreases with an increase in the number of applied cycles, corresponding to the left part of the hysteresis dissipated energy versus the interface shear-stress curve. The hysteresis loops correspond to the interface slip Case 4; that is, the interface completely debonds, and the fiber completely slides relative to the matrix in the interface debonding region on unloading/reloading.

1.4 HYSTERESIS MECHANISMS AND MODELS BASED ON EXPERIMENTAL OBSERVATIONS

1.4.1 Matrix Cracking Opening and Closure

For CMCs, matrix cracks can contract during unloading, even when the matrix is subject to residual tension, which is defined as matrix crack closure, as shown in Figure 1.28. Such behavior arises due to the occurrence of lateral grain-to-grain displacements as the matrix cracks form [9].

Figure 1.29a presents experimental and predicted matrix cracking densities, which were compared to the SiC/SiC minicomposite applied stress curves following loading/unloading. The SiC/SiC minicomposite is composed of an SiC fiber bundle, the SiC matrix surrounding the fibers, and the interface between the SiC fiber and the matrix. *In situ* an optical

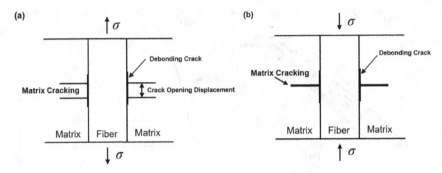

FIGURE 1.28 Schematic of matrix crack (a) opening and (b) closure of fiber-reinforced CMCs.

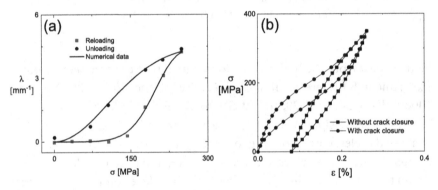

FIGURE 1.29 (a) Matrix crack density and (b) hysteresis loops without/with matrix crack closure of SiC/SiC composite.

microscope was used on the minicomposite to detect the evolution of matrix crack density with decreasing or increasing applied stress. With decreasing applied stress, the matrix cracking density decreases due to thermal residual compressive stress in the matrix, and with increasing applied stress, a matrix crack opening occurs again, leading to an increase of matrix crack density. However, evolution of matrix cracking density upon loading/unloading does not coincide with each other. Hysteresis loops were observed due to the fiber and matrix-induced frictional slips in the debonding regions following loading/unloading. Figure 1.29b shows loading/unloading hysteresis loops of SiC/SiC composite without/with matrix crack closure. It can be found that the shape, location, and area of the hysteresis loops are all affected by the matrix crack density.

The unloading hysteresis loops were categorized into two stages, specifically:

Stage I: Unloading stress, $\sigma_{\text{unloading}}$, had a value that was less than the peak stress, σ_{max}, and greater than the unloading transition stress, $\sigma_{\text{tr\_unloading}}$ (i.e., $\sigma_{\text{tr\_unloading}} < \sigma_{\text{unloading}} < \sigma_{\text{max}}$), wherein the interface reverse slip length approached the interface debonding length.

Stage II: Unloading stress, $\sigma_{\text{unloading}}$, was less than the unloading transition stress, $\sigma_{\text{tr\_unloading}}$, and greater than the valley stress, σ_{min} (i.e., $\sigma_{\text{min}} < \sigma_{\text{unloading}} < \sigma_{\text{tr\_unloading}}$).

In comparison, the reloading hysteresis loops were categorized into two stages, specifically:

Stage I: reloading stress, $\sigma_{\text{reloading}}$, was greater than the valley stress, σ_{min}, and less than the reloading transition stress, $\sigma_{\text{tr\_reloading}}$ (i.e., $\sigma_{\text{min}} < \sigma_{\text{reloading}} < \sigma_{\text{tr\_reloading}}$), wherein the interface new-slip length approached the interface debonding length.

Stage II: reloading stress, $\sigma_{\text{reloading}}$, was greater than the reloading transition stress, $\sigma_{\text{tr\_reloading}}$, and less than the peak stress, σ_{max} (i.e., $\sigma_{\text{tr\_reloading}} < \sigma_{\text{reloading}} < \sigma_{\text{max}}$).

Figure 1.30 shows the unloading and reloading axial stress distribution in the fiber and the matrix.

Figures 1.31 and 1.32 show the cyclic loading/unloading hysteresis loops, inverse tangent modulus (ITMs), interface reverse-slip ratio (IRSR), and interface new-slip ratio (INSR) without/with considering matrix crack closure for $V_f = 0.3$ and 0.35. Considering closure of matrix cracking, the peak strain of the hysteresis loops is the same as that without considering matrix crack closure; the residual strain of the hysteresis loops is less than that without considering matrix crack closure, due to the low matrix crack density upon unloading; and the cyclic dissipation of the hysteresis loops is higher than that without considering matrix crack closure, due to the difference of the matrix cracking density on loading/unloading. The evolution of the cyclic loading/unloading hysteresis dissipation, inverse tangent modulus (ITMs), IRSR, and INSR versus increasing/decreasing stress is much different from that without considering

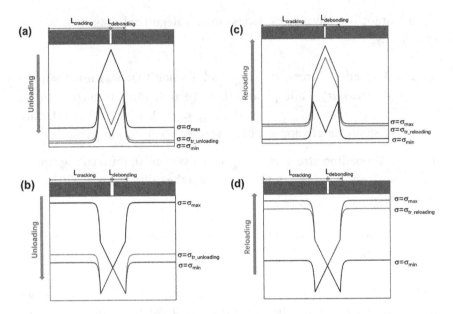

FIGURE 1.30   (a) Unloading fiber axial stress distribution, (b) unloading matrix axial stress distribution, (c) reloading fiber axial stress distribution, and (d) reloading fiber axial stress distribution.

matrix crack closure. Upon unloading, considering matrix crack closure, the unloading ITM increased to the peak value and then decreased, and the IRSR increased to the peak value and then decreased to zero; however, without considering matrix crack closure, the unloading ITM increased till the valley stress, and the IRSR increased to the peak value and then remained constant till the valley stress. Upon reloading, considering matrix crack closure, the reloading ITM nonlinear increased, and the INSR increased slowly at the initial stage of reloading, then increased to the peak value, and remained constant until peak stress; however, without considering matrix crack closure, the reloading ITM increased linearly, and the INSR increased to the peak value and then remained constant until the peak stress.

1.4.2 Interface Debonding and Slip

Interface debonding and slip affect the shape and location of mechanical hysteresis behavior of CMCs [10]. The shape and area of hysteresis loops of unidirectional SiC/CAS composite as a function of the interface shear stress are shown in Figure 1.33. The hysteresis loops of four cases

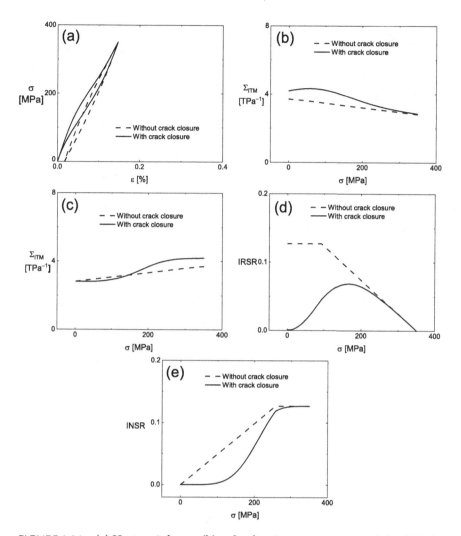

FIGURE 1.31 (a) Hysteresis loops, (b) unloading inverse tangent modulus (ITM), (c) reloading inverse tangent modulus (ITM), (d) unloading interface reverse-slip ratio (IRSR), and (e) reloading interface new-slip ratio (INSR) of SiC/SiC without/ with matrix crack closure for $V_f = 0.3$.

as a function of interface shear stress under $\sigma_{max} = 242$MPa are shown in Figure 1.33a, in which the shape and location are different from each other. The hysteresis dissipated energy as a function of interface shear stress ($\tau_i = 1\sim50$ MPa) is shown in Figure 1.33b. When $\tau_i = 27\sim50$ MPa, the hysteresis dissipated energy increases as interface shear stress decreases. The hysteresis loops correspond to Case 1; that is, the interface partially debonds,

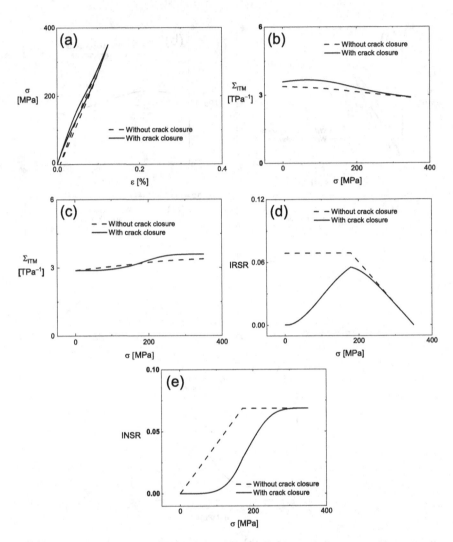

FIGURE 1.32 (a) Hysteresis loops, (b) unloading inverse tangent modulus (ITM), (c) reloading inverse tangent modulus (ITM), (d) unloading interface reverse-slip ratio (IRSR), and (e) reloading interface new-slip ratio (INSR) of SiC/SiC without/ with matrix crack closure for $V_f = 0.35$.

and the fiber completely slides relative to matrix in the debonding region. When $\tau_i = 8.8\sim27$ MPa, the hysteresis dissipated energy increases as the interface shear stress decreases. The hysteresis loops correspond to Case 2; that is, the interface partially debonds, and the fiber partially slides relative to the matrix in the interface debonding region. When $\tau_i = 8.3\sim8.8$

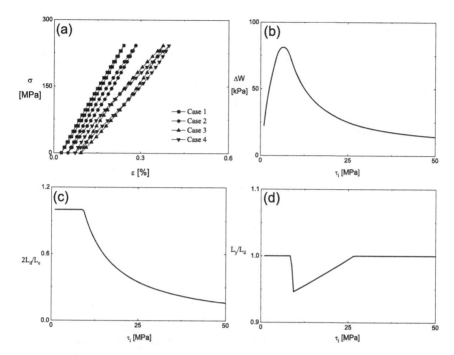

FIGURE 1.33 (a) Hysteresis loops, (b) hysteresis dissipated energy, (c) interface debonding length ($2L_d/L_c$), and (d) interface counter-slip length (L_y/L_d) of unidirectional SiC/CAS composite.

MPa, the hysteresis dissipated energy increases as the interface shear stress decreases. The hysteresis loops correspond to Case 3; that is, the interface completely debonds, and the fiber slides partially relative to the matrix in the debonding region. When $\tau_i = 1~8.3$ MPa, the hysteresis dissipated energy increase to the maximum and then decreases as interface shear stress decreases. The hysteresis loops correspond to Case 4; that is, the interface completely debonds, and the fiber slides completely relative to the matrix in the debonding region.

1.4.3 Fiber Failure and Pullout

Fiber failure and pullout affect the load distribution between intact and broken fibers and the interface debonding and slip in the debonding regions [11]. Experimental and predicted hysteresis loops of the unidirectional C/SiC composite with and without considering fiber failure under $\sigma_{max} = 260$ MPa are shown in Figure 1.34a. The fiber/matrix interface completely debonds, and the fiber failure volume fraction is P = 18.7%. The maximum

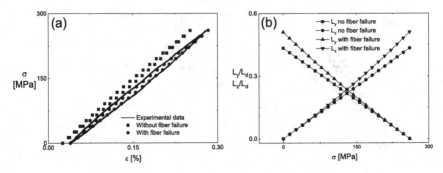

FIGURE 1.34 (a) Experimental and predicted hysteresis loops of unidirectional C/SiC composite and (b) interface counter-slip length L_y and reloading interface new-slip length L_z versus stress under σ_{max} = 260 MPa.

strain and residual strain, considering fiber failure, are obviously larger than those without consideration of fiber failure. The hysteresis loop with consideration of fiber failure agrees well with the experimental data. The unloading interface counter-slip length L_y and reloading interface new-slip length L_z as a function of applied stress for the present analysis with and without consideration of fiber failure are shown in Figure 1.34b. The unloading interface counter-slip length L_y increases as stress decreases until σ_{min}, at which the unloading interface counter-slip length L_y doesn't approach the half matrix crack spacing, $L_y(\sigma_{min}) < L_c/2$. The reloading interface new-slip length L_z increases as stress increases until σ_{max}, at which the reloading interface new-slip length L_z doesn't approach the half matrix crack spacing, $L_z(\sigma_{max}) < L_c/2$. The unloading interface counter-slip length and reloading new-slip length during unloading and subsequent reloading are obviously larger than those without consideration of fiber failure.

1.5 DISCUSSION

1.5.1 Effect of Temperature on Mechanical Hysteresis Behavior in CMCs

The tension-tension fatigue behavior of 2D woven SiC/SiC composite at 600, 800, and 1000°C in an inert atmosphere was investigated [12]. The fatigue peak stress was σ_{max} = 130 MPa, and the valley stress was σ_{min} = zero MPa. The loading frequency was f = 1 Hz. The fatigue hysteresis dissipated energy corresponding to different applied cycles at 600, 800, and 1000°C in an inert atmosphere is shown in Figure 1.35a. The fatigue hysteresis dissipated energy increases with the cycle number and, at the same cycle

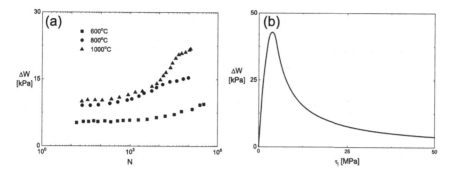

FIGURE 1.35 (a) Experimental fatigue hysteresis dissipated energy and (b) theoretical fatigue hysteresis dissipated energy of 2D SiC/SiC composite under σ_{max} = 130 MPa at T = 600, 800, and 1000°C.

number, increases with the test temperature. The interface shear stress corresponding to different applied cycles can be obtained from the hysteresis dissipated energy versus interface shear stress diagram, as shown in Figure 1.35b, by comparing the experimental fatigue hysteresis dissipated energy with theoretical computational values.

Experimental and theoretical interface shear stress as a function of cycle number at 600°C is shown in Figure 1.36a. The interface shear stress decreases from τ_i = 35 MPa at the 1st cycle to τ_i = 20.4 MPa at the 333507th cycle. The experimental and theoretical fatigue hysteresis dissipated energy versus cycle number curve is shown in Figure 1.36b. The fatigue hysteresis dissipated energy increases with the increasing cycle number from ΔW = 5.5 kPa at the 1st cycle to ΔW = 9.6 kPa at the 400000th cycle. The evolution of interface shear stress and fatigue hysteresis dissipated energy versus cycle number agreed with experimental data.

Experimental and theoretical interface shear stress as a function of cycle number at 800°C is shown in Figure 1.37a. The interface shear stress decreases from τ_i = 22 MPa at the 1st cycle to τ_i = 12.5 MPa at the 97894th cycle. The experimental and theoretical fatigue hysteresis dissipated energy versus cycle number curve is shown in Figure 1.37b. The fatigue hysteresis dissipated energy first increases with the increase of cycle number from ΔW = 9.2 kPa at the 1st cycle to the peak value ΔW = 15.4 kPa at the 122364th cycle and then remains to be constant to 400000th cycle. Evolution of interface shear stress and fatigue hysteresis dissipated energy versus cycle number agreed with experimental data.

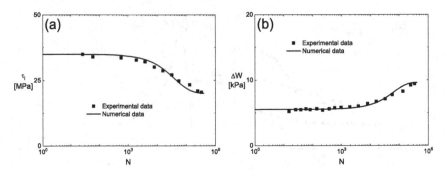

FIGURE 1.36   (a) Interface shear stress, and (b) fatigue hysteresis dissipated energy of 2D SiC/SiC composite under σ_{max} = 130 MPa at 600°C in an inert atmosphere.

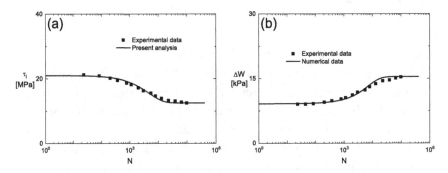

FIGURE 1.37   (a) Interface shear stress and (b) fatigue hysteresis dissipated energy of 2D SiC/SiC composite under σ_{max} = 130 MPa at 800°C in an inert atmosphere.

Experimental and theoretical interface shear stress as a function of cycle number at 1000°C is shown in Figure 1.38a. The interface shear stress decreases from τ_i = 18 MPa at the 1st cycle to τ_i = 8.5 MPa at the 117055th cycle. The experimental and theoretical fatigue hysteresis dissipated energy versus cycle number curve is shown in Figure 1.38b. The fatigue hysteresis dissipated energy increases with increasing cycle number from ΔW = 10.7 kPa at the 1st cycle to the peak value ΔW = 22.6 kPa at the 375365th cycle and then remains to be constant to 400000th cycle. Evolution of interface shear stress and fatigue hysteresis dissipated energy versus cycle number agreed with experimental data.

1.5.2 Effect of Loading Frequency on Mechanical Hysteresis Behavior in CMCs

The effect of loading frequency on the fatigue hysteresis dissipated energy versus cycle number curves of cross-ply SiC/MAS composite at 566°C and

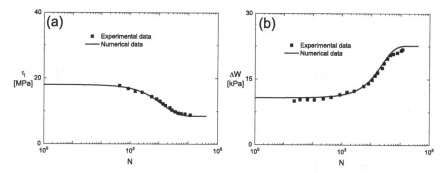

FIGURE 1.38   (a) Interface shear stress and (b) fatigue hysteresis dissipated energy of 2D SiC/SiC composite under σ_{max} = 130 MPa at 1000°C in an inert atmosphere.

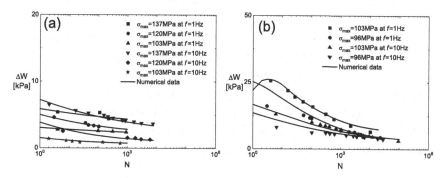

FIGURE 1.39   Effect of loading frequency on fatigue hysteresis dissipated energy versus cycle number curves of cross-ply SiC/MAS composite at (a) T = 566°C in air and (b) T = 1093°C in air.

1093°C in air is shown in Figure 1.39. The fatigue hysteresis dissipated energy at low loading frequency is higher than that at high loading frequency; and the fatigue hysteresis dissipated energy degradation rate at low loading frequency is also higher than that at high loading frequency.

At 566°C in an air atmosphere under σ_{max} = 137 MPa, the fatigue hysteresis dissipated energy decreases from ΔW = 5.4 kPa at the 4th cycle to ΔW = 4.4 kPa at the 230th cycle, and the fatigue hysteresis dissipated energy degradation rate is 4.4 × 10^{-3} kPa/cycle with the loading frequency f = 1 Hz; and when f = 10 Hz, the fatigue hysteresis dissipated energy decreases from ΔW = 6.5 kPa at the 2nd cycle to ΔW = 3.6 kPa at the 7730th applied cycle, and the fatigue hysteresis dissipated energy degradation rate is 3.7 × 10^{-4} kPa/cycle. Under σ_{max} = 120 MPa, the fatigue hysteresis dissipated energy decreases from ΔW = 4.6 kPa at the 3rd applied cycle to ΔW = 3.2 kPa at the 105th applied cycle, and the fatigue hysteresis dissipated energy degradation rate is 1.3×10^{-2} kPa/cycle with the loading frequency $f = 1$ Hz; and when $f = 10$ Hz, the fatigue hysteresis dissipated energy decreases from $\Delta W = 2.5$ kPa at the 6th applied cycle to $\Delta W = 1.3$ kPa at the 6150th applied cycle, and the fatigue hysteresis dissipated energy degradation rate is 1.9×10^{-4} kPa/cycle.

At 1093°C in air atmosphere under $\sigma_{max} = 103$ MPa, the fatigue hysteresis dissipated energy decreases from $\Delta W = 25.5$ kPa at the 4th cycle to $\Delta W = 6.5$ kPa at the 10608th applied cycle, and the fatigue hysteresis dissipated energy degradation rate is 1.8×10^{-3} kPa/cycle; and at the loading frequency $f = 10$ Hz, the fatigue hysteresis dissipated energy decreases from $\Delta W = 13$ kPa at the 6th applied cycle to $\Delta W = 3.1$ kPa at the 94044th applied cycle, and the fatigue hysteresis dissipated energy degradation rate is 1.0×10^{-4} kPa/cycle. Under $\sigma_{max} = 96$ MPa, the fatigue hysteresis dissipated energy decreases from $\Delta W = 16$ kPa at the 3rd applied cycle to $\Delta W = 4.4$ kPa at the 23067th applied cycle, and the fatigue hysteresis dissipated energy degradation rate is 5.0×10^{-4} kPa/cycle, with the loading frequency $f = 1$ Hz; and when $f = 10$ Hz, the fatigue hysteresis dissipated energy decreases from $\Delta W = 8.1$ kPa at the 6th applied cycle to $\Delta W = 3.6$ kPa at the 28400th applied cycle, and the fatigue hysteresis dissipated energy degradation rate is 1.5×10^{-4} kPa/cycle.

1.5.3 Effect of Fatigue Stress Ratio on Mechanical Hysteresis Behavior in CMCs

The effect of stress ratio, that is, $R = 0, 0.1$, and 0.2, on hysteresis dissipated energy of SiC/CAS composite under $\sigma_{max} = 210, 220, 240, 300$, and 320 MPa are shown in Figure 1.40. With the increase of stress ratio, the range and extent of interface frictional slip between fibers and the matrix in the interface debonding region would decrease. Hysteresis dissipated energy decreases with the increase of stress ratio at the same cycle number.

- Under $\sigma_{max} = 210$ MPa, hysteresis dissipated energy decreases with the increase of stress ratio, that is, from $\Delta W = 4.7$ kPa at the 1st cycle to $\Delta W = 46.9$ kPa at the 2577th cycle first, then decreases to $\Delta W = 29.5$ kPa at the 4784th cycle, and increases to $\Delta W = 36.3$ kPa at the 20370th cycle for $R = 0$; and hysteresis dissipated energy increases from $\Delta W = 2.4$ kPa at the 1st cycle to $\Delta W = 30$ kPa at the 2797th cycle, then decreases to $\Delta W = 20.2$ kPa at the 5170th cycle, and increases again to $\Delta W = 23.3$ kPa at the 20370th cycle for $R = 0.2$.

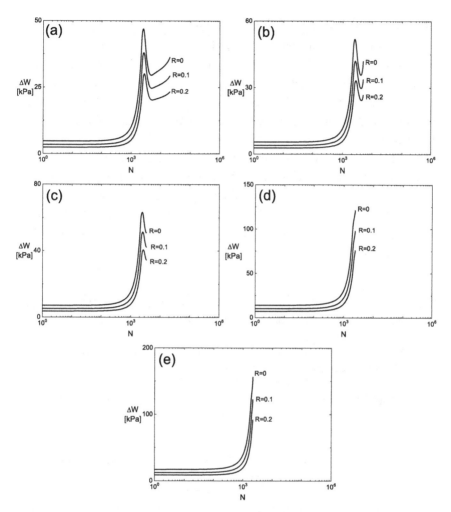

FIGURE 1.40 Effect of fatigue stress ratio, that is, $R = 0$, 0.1 and 0.2, on the fatigue hysteresis dissipated energy versus cycle number curve for different fatigue peak stresses (a) $\sigma_{max} = 210$ MPa, (b) $\sigma_{max} = 220$ MPa, (c) $\sigma_{max} = 240$ MPa, (d) $\sigma_{max} = 300$ MPa, and (e) $\sigma_{max} = 320$ MPa.

- Under $\sigma_{max} = 320$ MPa, hysteresis dissipated energy decreases with the increase of stress ratio, that is, from $\Delta W = 16.9$ kPa at the 1st cycle to $\Delta W = 156$ kPa at the 2171th cycle for $R = 0$; and hysteresis dissipated energy increases from $\Delta W = 8.6$ kPa at the 1st cycle to $\Delta W = 91.7$ kPa at the 2171th cycle for $R = 0.2$.

1.6 SUMMARY AND CONCLUSION

In this chapter, the application background of CMCs on aircraft or aero-engines was presented. The manufacturing of CMCs, including fibers, fabric architecture, interface, and matrix, is introduced. The mechanical properties of CMCs fabricated by the CVI, PIP, MI and HP methods were analyzed. High-temperature mechanical hysteresis behavior of unidirectional C/SiC, cross-ply C/SiC and SiC/MAS-L, 2D plain-woven SiC/SiC, and 2.5D C/SiC were compared and discussed. Under cyclic loading/unloading, damage mechanisms of matrix crack opening and closure, interface debonding and slip, and fiber failure and pullout contribute to the hysteresis appearance of CMCs. The effects of temperature, loading frequency, and stress ratio on mechanical hysteresis behavior of CMCs were discussed.

REFERENCES

1. Naslain R. Design, preparation and properties of non-oxide CMCs for application in engines and nuclear reactors: an overview. *Compos. Sci. Technol.* 2004; 64(2):155–170.
2. DiCarlo JA, Van Roode M. Ceramic composite development for gas turbine hot section components. *Proceedings of the ASME Turbo Expo: Power for Land, Sea and Air,* 2006; 2:221–231.
3. Zhang ZW, Li LB, Chen ZK. Damage evolution and fracture behavior of C/SiC minicomposites with different interphases under uniaxial tensile load. *Materials* 2021; 14:1525.
4. Yang HT, Xu SW, Zhang DX, Li LB, Huang XZ. In-situ tensile damage and fracture behavior of PIP SiC/SiC minicomposites at room temperature. *J. Eur. Ceram. Soc.* 2021; 41:6869–6882.
5. Liu H, Li LB, Yang JH, Zhou Y, Ai Y, Qi Z, Gao Y, Jiao J. Characterization and modeling damage and fracture of Prepreg-MI SiC/SiC composite under tensile loading at room temperature. *Appl. Compos. Mater.* 2022. https://doi.org/10.1007/s10443-022-10015-6
6. Li LB, Song YD, Sun YC. Modeling the tensile behavior of unidirectional C/SiC ceramic matrix composites. *Mech. Compos. Mater.* 2014; 49:659–672.
7. Li LB, Song YD, Sun YC. Modeling tensile behavior of cross-ply C/SiC ceramic-matrix composites. *Mech. Compos. Mater.* 2015; 51:358–376.
8. Li LB. Fatigue hysteresis of carbon fiber-reinforced ceramic-matrix composites at room and elevated temperatures. *Appl. Compos. Mater.* 2016; 23:1–27.
9. Li LB. A micromechanical loading/unloading constitutive model of fiber-reinforced ceramic-matrix composites considering matrix crack closure. *Fatigue Fract. Eng. Mater. Struct.* 2021; 44:2389–2411.

10. Li LB, Song YD. An approach to estimate interface shear stress of ceramic matrix composites from hysteresis loops. *Appl. Compos. Mater.* 2010; 17:309–328.

11. Li LB, Song YD, Sun YC. Modeling loading/unloading hysteresis behavior of unidirectional C/SiC ceramic matrix composites. *Appl. Compos. Mater.* 2013; 20:655–672.

12. Li LB. Effects of temperature, oxidation and fiber preforms on interface shear stress degradation in fiber-reinforced ceramic-matrix composites. *Mater. Sci. Eng. A* 2016; 674:588–603.

CHAPTER 2

Cyclic Mechanical Hysteresis Behavior in One-Dimensional SiC/SiC Minicomposites at Room Temperature

2.1 INTRODUCTION

Critical to the success and further implementation of fiber-reinforced ceramic-matrix composites (CMCs) is a thorough understanding of the nature of damage development under tensile or cyclic loading. Cho et al. [1] and Li [2] established the relationships between surface temperature rising, hysteresis dissipated energy, and internal damage of CMCs. Fantozzi and Reynaud [3], Mei and Cheng [4], and Li [5] analyzed the effect of fiber performance and fiber volume along the loading direction on hysteresis loops of CMCs. However, due to the complex damage mechanisms in woven structures, the minicomposites are usually used to study the damage mechanisms within the woven tows, which affects the mechanical behavior of woven CMCs [6–10]. Maillet et al. [11], Almansour et al. [12], and Zhang et al. [13] found that the damage and fracture of SiC/SiC minicomposites related to the type of fiber and the interphase.

In this chapter, the cyclic mechanical hysteresis behavior of SiC/SiC minicomposites with different fiber types and interface properties is investigated. The synergistic effects of interface debonding and slip and fiber

DOI: 10.1201/b23026-2

fracture on stress-dependent mechanical hysteresis of SiC/SiC minicomposites are analyzed. The relationships among the stress, interface debonding and slip, fiber broken, and mechanical hysteresis are established for SiC/SiC minicomposite. Effects of constituent properties and internal damage state on the cyclic loading/unloading interface slip and the hysteresis loops of SiC/SiC minicomposite are analyzed. The experimental matrix cracking evolution, interface debonding and slip, and the hysteresis loops of Hi-Nicalon™, Hi-Nicalon™ Type S, and Tyranno™ ZMI SiC/SiC minicomposites are predicted.

2.2 MICROMECHANICAL HYSTERESIS CONSTITUTIVE MODEL

Under cyclic loading/unloading, hysteresis loops appear in CMCs mainly due to the interface slip in the debonding regions [14–21]. The state of interface debonding and slip affects mechanical hysteresis of mini-CMCs. When fiber fracture occurs inside of CMCs, the interface shear stress transfers load between the fracture fibers and the matrix, and the global load sharing (GLS) criterion is used to determine the loading distribution between the fiber and the matrix. When the interface partial debonds, the unloading/reloading hysteresis strain ε_U and ε_R are

$$
\begin{aligned}
\varepsilon_U(\sigma) = & \frac{\Phi_U}{E_f} + 4\frac{\tau_i}{E_f}\frac{L_y^2}{r_f L_c(\sigma_{max})} - \frac{\tau_i}{r_f E_f L_c(\sigma_{max})} \\
& \times \left[2L_y - L_d(\sigma_{max})\right]\left[2L_y + L_d(\sigma_{max}) - L_c(\sigma_{max})\right] \\
& - (\alpha_c - \alpha_f)\Delta T,
\end{aligned}
\tag{2.1}
$$

$$
\begin{aligned}
\varepsilon_R(\sigma) = & \frac{\Phi_R}{E_f} - 4\frac{\tau_i}{E_f}\frac{L_z^2(\sigma)}{r_f l_c(\sigma_{max})} + 4\frac{\tau_i}{E_f}\frac{\left[L_y(\sigma_{min}) - 2L_z(\sigma)\right]^2}{r_f l_c(\sigma_{max})} \\
& + 2\frac{\tau_i}{r_f E_f L_c}\left[L_d(\sigma_{max}) - 2L_y(\sigma_{min}) + 2L_z(\sigma)\right] \\
& \times \left[L_d(\sigma_{max}) + 2L_y(\sigma_{min}) - 2L_z(\sigma) - L_c(\sigma_{max})\right] \\
& - (\alpha_c - \alpha_f)\Delta T,
\end{aligned}
\tag{2.2}
$$

where Φ_U and Φ_R denote the unloading and reloading of intact fiber stress, respectively; E_f denotes fiber elastic modulus; r_f denotes fiber radius;

τ_i denotes interface shear stress; α_f and α_c denote fiber and composite thermal expansion coefficient, respectively; ΔT denotes temperature difference between fabricated temperature and test temperature; L_d and L_c denote the interface debonding length and matrix crack spacing, respectively; and L_y and L_z denote the unloading/reloading interface slip length, respectively.

$$\frac{\sigma}{V_f} = 2\Phi\left(\frac{\sigma_{fc}}{\Phi}\right)^{m_f+1}\left\{\exp\left[-\left(\frac{\Phi-\Phi_U}{2\Phi}\right)\left(\frac{\Phi}{\sigma_{fc}}\right)^{m_f+1}\right]-1+\frac{1}{2}P(\Phi)\right\}, \quad (2.3)$$

$$\frac{\sigma}{V_f} = 2\Phi\left(\frac{\sigma_{fc}}{\Phi}\right)^{m_f+1}\left\{\exp\left[-\left(\frac{\Phi_m}{2\Phi}\right)\left(\frac{\Phi}{\sigma_{fc}}\right)^{m_f+1}\right]\right.$$
$$\left.-\exp\left[-\left(\frac{\Phi_R-\Phi+\Phi_m}{2\Phi}\right)\left(\frac{\Phi}{\sigma_{fc}}\right)^{m_f+1}\right]+\frac{1}{2}P(\Phi)\right\}, \quad (2.4)$$

where V_f denotes the fiber volume, Φ denotes the intact fiber stress at the peak stress, m_f denotes the fiber Weibull modulus, $P(\Phi)$ denotes the fiber failure probability, and σ_{fc} denotes the fiber characteristic strength.

When the interface complete debonds, the unloading/reloading hysteresis strain ε_U and ε_R are

$$\varepsilon_U(\sigma) = \frac{\Phi_U}{E_f} + 4\frac{\tau_i}{E_f}\frac{L_y^2(\sigma)}{r_f L_c(\sigma_{max})}$$
$$-2\frac{\tau_i}{E_f}\frac{(2L_y(\sigma)-L_c(\sigma_{max})/2)^2}{r_f L_c(\sigma_{max})}-(\alpha_c-\alpha_f)\Delta T, \quad (2.5)$$

$$\varepsilon_R(\sigma) = \frac{\Phi_R}{E_f} - 4\frac{\tau_i}{E_f}\frac{L_z^2(\sigma)}{r_f L_c(\sigma_{max})}+4\frac{\tau_i}{E_f}\frac{\left[L_y(\sigma_{min})-2L_z(\sigma)\right]^2}{r_f l_c(\sigma_{max})}$$
$$-2\frac{\tau_i}{E_f}\frac{\left[L_c(\sigma_{max})/2-2L_y(\sigma_{min})+2L_z(\sigma)\right]^2}{r_f L_c(\sigma_{max})}-(\alpha_c-\alpha_f)\Delta T. \quad (2.6)$$

Based on the interface debonding and slip state, the hysteresis loops of CMCs can be divided into four different cases:

Case I, partial interface debonding ($L_d < L_c/2$) and complete sliding ($L_y = L_z = L_d$)

Case II, partial interface debonding ($L_d < L_c/2$) and partial sliding ($L_y = L_z < L_d$)

Case III, complete interface debonding ($L_d = L_c/2$) and partial sliding ($L_y = L_z < L_c/2$)

Case IV, complete interface debonding ($L_d = L_c/2$), and complete sliding ($L_y = L_z = L_c/2$)

2.3 EXPERIMENTAL COMPARISONS

Almansour et al. [12] investigated cyclic loading/unloading tensile hysteresis behavior of single-tow SiC/SiC minicomposite from a single tow of SiC fibers of one of the following types: Hi-Nicalon™ (HN; Nippon Carbon, Tokyo, Japan), Hi-Nicalon™ Type S (HNS; Nippon Carbon, Tokyo, Japan), and Tyranno™ ZMI (ZMI; Ube Industry Led., Tokyo, Japan). The fibers were coated with a boron nitride (BN) or a carbon (C) interphase deposited by chemical vapor infiltration (CVI). The SiC matrix was then also deposited by CVI. The materials properties of mini-SiC/SiC composites are shown in Table 2.1.

TABLE 2.1   Material Properties of SiC/SiC Minicomposites

Items	Hi-Nicalon™ SiC/SiC	Hi-Nicalon™ Type S SiC/SiC	Tyranno™ ZMI SiC/SiC
$r_f/(\mu m)$	7	6	5.5
$V_f/(\%)$	25.8	22.8	27.5
$E_f/(GPa)$	270	400	170
$E_m/(GPa)$	350	350	350
$\alpha_f/(10^{-6}/°C)$	3.5	4.5	4.0
$\alpha_m/(10^{-6}/°C)$	4.6	4.6	4.6
$\sigma_R/(MPa)$	420	350	280
m	6	6	8
$L_{sat}/(\mu m)$	564	667	667
$\sigma_0/(GPa)$	2.4	3.0	1.9
m_f	5	5	5

2.3.1 Hi-Nicalon™ SiC/SiC Minicomposite

For the Hi-Nicalon™ SiC/SiC minicomposite, the experimental and predicted matrix cracking density, mechanical hysteresis loops, and interface slip are shown in Figures 2.1 and 2.2.

- The stress range for the matrix cracking evolution is between the first matrix cracking stress σ_{mc} = 250 MPa and the saturation matrix cracking stress σ_{sat} = 580 MPa, and the saturation matrix cracking density is λ_{sat} = 1.7/mm.

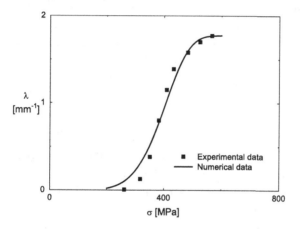

FIGURE 2.1 Experimental and predicted matrix cracking density of Hi-Nicalon™ SiC/SiC minicomposite.

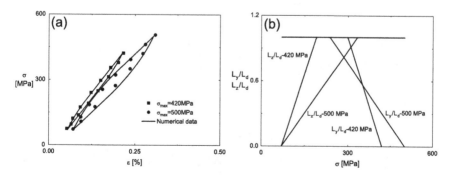

FIGURE 2.2 (a) Hysteresis loops and (b) interface slip length of Hi-Nicalon™ SiC/SiC minicomposite under σ_{max} = 420 and 500 MPa.

- Under σ_{max} = 420 MPa, the unloading transition stress is $\sigma_{tr\_unloading}$ = 297 MPa (i.e., $L_y(\sigma_{tr\_unloading}) = L_d(\sigma_{max})$), the reloading transition stress is $\sigma_{tr\_reloading}$ = 193 MPa (i.e., $L_z(\sigma_{tr\_reloading}) = L_d(\sigma_{max})$), and the mechanical hysteresis loops correspond to Case I; that is, the interface partially debonds (i.e., $2L_d/L_c < 1$), and the fiber slides complete relative to the matrix (i.e., $L_y/L_d = 1$) in the interface debonding region upon unloading and reloading.

- Under σ_{max} = 500 MPa, the unloading transition stress is $\sigma_{tr\_unloading}$ = 223 MPa (i.e., $L_y(\sigma_{tr\_unloading}) = L_d(\sigma_{max})$), the reloading transition stress is $\sigma_{tr\_reloading}$ = 337 MPa (i.e., $L_z(\sigma_{tr\_reloading}) = L_d(\sigma_{max})$), and the mechanical hysteresis loops correspond to Case I; that is, the interface partial debonds (i.e., $2L_d/L_c < 1$) and the fiber slides complete relative to the matrix (i.e., $L_y/L_d = 1$) in the interface debonding region on unloading and reloading.

2.3.2 Hi-Nicalon™ Type S SiC/SiC Minicomposite

For the Hi-Nicalon™ Type S SiC/SiC minicomposite, the experimental and predicted matrix cracking density, mechanical hysteresis loops, and interface slip are shown in Figures 2.3 and 2.4.

The matrix cracking stress range is between the first matrix cracking stress σ_{mc} = 220 MPa and saturation matrix cracking stress σ_{sat} = 470 MPa, and the saturation matrix cracking density is λ_{sat} = 1.5/mm.

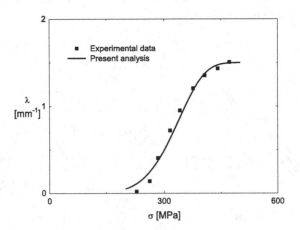

FIGURE 2.3 Experimental and predicted matrix cracking density of Hi-Nicalon™ Type S SiC/SiC minicomposite.

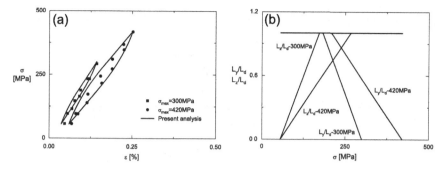

FIGURE 2.4 (a) Hysteresis loops and (b) interface slip length of Hi-Nicalon™ Type S SiC/SiC minicomposite under σ_{max} = 300 and 420 MPa.

Under σ_{max} = 300 MPa, the unloading transition stress is $\sigma_{tr\_unloading}$ = 182 MPa (i.e., $L_y(\sigma_{tr\_unloading}) = L_d(\sigma_{max})$), the reloading transition stress is $\sigma_{tr\_reloading}$ = 172 MPa (i.e., $L_z(\sigma_{tr\_reloading}) = L_d(\sigma_{max})$), and the mechanical hysteresis loops correspond to Case I; that is, the interface partial debonds (i.e., $2L_d/L_c < 1$) and the fiber slides complete relative to the matrix (i.e., $L_y(L_z)/L_d = 1$) in the interface debonding region upon unloading and reloading.

Under σ_{max} = 420 MPa (i.e., $L_y(\sigma_{tr\_unloading}) = L_d(\sigma_{max})$), the unloading transition stress is $\sigma_{tr\_unloading}$ = 208 MPa, the reloading transition stress is $\sigma_{tr\_reloading}$ = 267 MPa (i.e., $L_z(\sigma_{tr\_reloading}) = L_d(\sigma_{max})$), and the mechanical hysteresis loops correspond to Case I; that is, the interface partial debonds (i.e., $2L_d/L_c < 1$) and the fiber slides complete relative to the matrix (i.e., $L_y(L_z)/L_d = 1$) in the interface debonding region on unloading and reloading.

2.3.3 Tyranno™ ZMI SiC/SiC Minicomposite

For the Tyranno™ ZMI SiC/SiC minicomposite, the experimental and predicted matrix cracking density, mechanical hysteresis loops, and interface slip are shown in Figures 2.5 and 2.6.

The matrix cracking stress range is between the first matrix cracking stress σ_{mc} = 150 MPa and the saturation matrix cracking stress σ_{sat} = 370 MPa, and the saturation matrix cracking density is λ_{sat} = 1.5/mm.

Under σ_{max} = 300 MPa, the unloading transition stress is $\sigma_{tr\_unloading}$ = 274 MPa (i.e., $L_y(\sigma_{tr\_unloading}) = L_d(\sigma_{max})$), the reloading transition stress is $\sigma_{tr\_reloading}$ = 246 MPa (i.e., $L_z(\sigma_{tr\_reloading}) = L_d(\sigma_{max})$), and the mechanical hysteresis loops correspond to Case I; that is, the interface partially

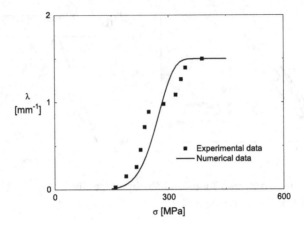

FIGURE 2.5   Experimental and predicted matrix cracking density of Tyranno™ ZMI SiC/SiC minicomposite.

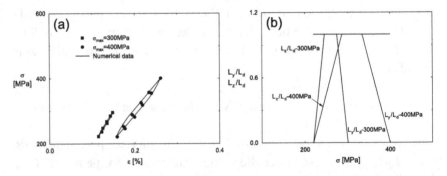

FIGURE 2.6   (a) Hysteresis loops and (b) interface slip length of Tyranno™ ZMI SiC/SiC minicomposite under $\sigma_{max} = 300$ and 400 MPa.

debonds (i.e., $2L_d/L_c < 1$), and the fiber slides completely relative to the matrix (i.e., $L_y(L_z)/l_d = 1$) in the interface debonding region on unloading and reloading.

Under $\sigma_{max} = 400$ MPa, the unloading transition stress is $\sigma_{tr\_unloading} = 333$ MPa (i.e., $L_y(\sigma_{tr\_unloading}) = L_d(\sigma_{max})$), the reloading transition stress is $\sigma_{tr\_reloading} = 287$ MPa (i.e., $L_z(\sigma_{tr\_reloading}) = L_d(\sigma_{max})$), and the mechanical hysteresis loops correspond to Case I; that is, the interface partially debonds (i.e., $2L_d/L_c < 1$) and the fiber slides completely relative to the matrix (i.e., $L_y(L_z)/L_d = 1$) in the interface debonding region on unloading and reloading.

2.4 DISCUSSIONS

The Hi-Nicalon™ SiC/SiC minicomposite is used for the case analysis. Effects of fiber volume fraction, interface shear stress, interface debonding energy, matrix crack spacing, and fiber failure on mechanical hysteresis loops are analyzed.

2.4.1 Effect of Fiber Volume Fraction on Mechanical Hysteresis Loops

The mechanical hysteresis loops and interface slip of mini-SiC/SiC composite under σ_{max} = 350, 400, and 450 MPa for V_f = 25% and 30% are shown in Figure 2.7. When the fiber volume fraction increases, the interface debonding and slip lengths decrease, the hysteresis loops area and peak strain decrease, and the hysteresis modulus increases.

When V_f = 25% at the peak stress σ_{max} = 350 MPa, the interface debonding length occupies 52.8% of the matrix crack spacing, that is, $2L_d(\sigma_{max}$ = 350 MPa)/L_c = 52.8%; the interface counter-slip length approaches to the interface debonding length at the unloading stress of $\sigma_{tr\_unloading}$ = 65 MPa,

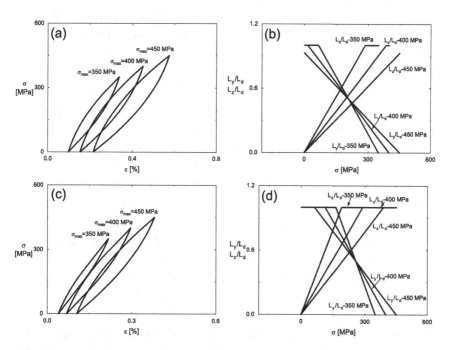

FIGURE 2.7 (a) Hysteresis loops when V_f = 25%, (b) interface slip length when V_f = 25%; (c) hysteresis loops when V_f = 30%, and (d) interface slip length when V_f = 30%.

that is, $L_y(\sigma_{tr\_unloading} = 65$ MPa$)/L_d = 1.0$; the interface new-slip length approaches to the interface debonding length at the reloading stress $\sigma_{tr\_reloading} = 285$ MPa, that is, $L_z(\sigma_{tr\_reloading} = 285$ MPa$)/L_d = 1.0$; and the hysteresis loops correspond to the interface slip Case 1. Under $\sigma_{max} = 400$ MPa, the interface debonding length occupies 71.4% of the matrix crack spacing, that is, $2L_d(\sigma_{max} = 400$ MPa$)/L_c = 71.4\%$; the interface counter-slip length approaches to the interface debonding length at the unloading stress $\sigma_{tr\_unloading} = 15$ MPa, that is, $L_y(\sigma_{tr\_unloading} = 15$ MPa$)/L_d = 1.0$; the interface new-slip length approaches to the interface debonding length at the reloading stress $\sigma_{tr\_reloading} = 385$ MPa, that is, $L_z(\sigma_{tr\_reloading} = 385$ MPa$)/L_d = 1.0$; and the hysteresis loops correspond to the interface slip Case 1. Under $\sigma_{max} = 450$ MPa, the interface debonding length occupies 89.9% of the matrix crack spacing, that is, $2L_d(\sigma_{max} = 450$ MPa$)/L_c = 89.9\%$; the interface counter-slip length occupies 92.8% of the interface debonding length, that is, $L_y(\sigma_{min})/L_d = 92.8\%$; the interface new-slip length occupies 92.8% of the interface debonding length, that is, $L_z(\sigma_{max})/L_d = 92.8\%$; and the hysteresis loops correspond to the interface slip Case 2.

When $V_f = 30\%$ at the peak stress $\sigma_{max} = 350$ MPa, the interface debonding length occupies 27.5% of the matrix crack spacing, that is, $2L_d(\sigma_{max} = 350$ MPa$)/L_c = 27.5\%$; the interface counter-slip length approaches to the interface debonding length at the unloading stress $\sigma_{tr\_unloading} = 161$ MPa, that is, $L_y(\sigma_{tr\_unloading} = 162$ MPa$)/L_d = 1.0$; the interface new-slip length approaches to the interface debonding length at the reloading stress $\sigma_{tr\_reloading} = 188$ MPa, i.e., $L_z(\sigma_{tr\_reloading} = 188$ MPa$)/L_d = 1.0$; and the hysteresis loops correspond to the interface slip Case 1. Under $\sigma_{max} = 400$ MPa, the interface debonding length occupies 42.1% of the matrix crack spacing, that is, $2L_d(\sigma_{max} = 400$ MPa$)/L_c = 42.1\%$; the interface counter-slip length approaches to the interface debonding length at the unloading stress $\sigma_{tr\_unloading} = 112$ MPa, that is, $L_y(\sigma_{tr\_unloading} = 112$ MPa$)/L_d = 1$; the interface new-slip length approaches to the interface debonding length at the reloading stress $\sigma_{tr\_reloading} = 288$ MPa, that is, $L_z(\sigma_{tr\_reloading} = 288$ MPa$)/L_d = 1$; and the hysteresis loops correspond to the interface slip Case 1. Under $\sigma_{max} = 450$ MPa, the interface debonding length occupies 56.7% of the matrix crack spacing, that is, $2L_d(\sigma_{max} = 450$ MPa$)/L_c = 56.7\%$; the interface counter-slip length approaches to the interface debonding length at the unloading stress $\sigma_{tr\_unloading} = 62$ MPa, that is, $L_y(\sigma_{tr\_unloading} = 62$ MPa$)/L_d = 1.0$; the interface new-slip length approaches to the interface debonding length at the reloading stress $\sigma_{tr\_reloading}$ = 388 MPa, that is, $L_z(\sigma_{tr\_reloading}$ = 388 MPa)/L_d = 1.0; and the hysteresis loops correspond to the interface slip Case 1.

2.4.2 Effect of Interface Shear Stress on Mechanical Hysteresis Loops

The hysteresis loops and interface slip of mini-SiC/SiC composite under σ_{max} = 350, 400, and 450 MPa for τ_i = 40 and 50 MPa are shown in Figure 2.8. When the interface shear stress increases, the interface debonding and slip length decrease, the hysteresis loops area and peak strain decrease, and the hysteresis modulus increases.

When τ_i = 40 MPa at the peak stress σ_{max} = 350 MPa, the interface debonding length occupies 39.2% of the matrix crack spacing, i.e., $2L_d(\sigma_{max}$ = 350 MPa)/L_c = 39.2%; the interface counter-slip length approaches to the interface debonding length at the unloading stress $\sigma_{tr\_unloading}$ = 68 MPa, that is, $L_y(\sigma_{tr\_unloading}$ = 68 MPa)/L_d = 1.0; the interface new-slip length

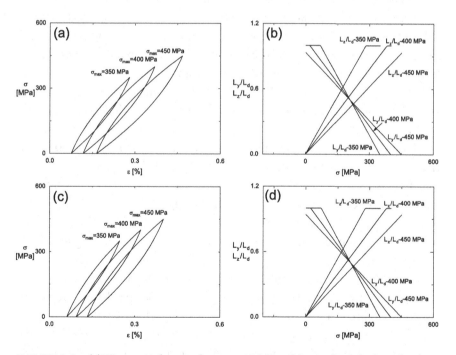

FIGURE 2.8 (a) Hysteresis loops when τ_i = 40 MPa, (b) interface slip length when τ_i = 40 MPa, (c) hysteresis loops when τ_i = 50 MPa, and (d) interface slip length when τ_i = 50 MPa.

approaches to the interface debonding length at the reloading stress $\sigma_{tr\_reloading} = 282$ MPa, i.e., $L_z(\sigma_{tr\_reloading} = 282$ MPa$)/L_d = 1.0$; and the hysteresis loops correspond to the interface slip Case 1. Under $\sigma_{max} = 400$ MPa, the interface debonding length occupies 53.1% of the matrix crack spacing, that is, $2L_d(\sigma_{max} = 400$ MPa$)/L_c = 53.1\%$; the interface counter-slip length approaches to the interface debonding length at the unloading stress $\sigma_{tr\_unloading} = 18$ MPa, that is, $L_y(\sigma_{tr\_unloading} = 18$ MPa$)/L_d = 1.0$; the interface new-slip length approaches to the interface debonding length at the reloading stress $\sigma_{tr\_reloading} = 382$ MPa, that is, $L_z(\sigma_{tr\_reloading} = 382$ MPa$)/L_c = 1.0$; and the hysteresis loops correspond to the interface slip Case 1. Under $\sigma_{max} = 450$ MPa, the interface debonding length occupies 67% of the matrix crack spacing, that is, $2L_d(\sigma_{max} = 450$ MPa$)/L_c = 67\%$; the interface counter-slip length approaches to 93% of the interface debonding length, that is, $L_y(\sigma_{min})/L_d = 93\%$; the interface new-slip length approaches to 93% of the interface debonding length, that is, $L_z(\sigma_{max})/L_d = 93\%$; and the hysteresis loops correspond to the interface slip Case 2.

When $\tau_i = 50$ MPa at the peak stress $\sigma_{max} = 350$ MPa, the interface debonding length occupies 31% of the matrix crack spacing, that is, $2L_d(\sigma_{max} = 350$ MPa$)/L_c = 31\%$; the interface counter-slip length approaches to the interface debonding length at the unloading stress $\sigma_{tr\_unloading} = 71$ MPa, that is, $L_y(\sigma_{tr\_unloading} = 71$ MPa$)/L_d = 1.0$; the interface new-slip length approaches to the interface debonding length at the reloading stress $\sigma_{tr\_reloading} = 279$ MPa, that is, $L_z(\sigma_{tr\_reloading} = 279$ MPa$)/L_d = 1.0$; and the hysteresis loops correspond to the interface slip Case 1. Under $\sigma_{max} = 400$ MPa, the interface debonding length occupies 42.1% of the matrix crack spacing, that is, $2L_d(\sigma_{max} = 400$ MPa$)/L_c = 42.1\%$; the interface counter-slip length approaches to the interface debonding length at the unloading stress $\sigma_{tr\_unloading} = 21$ MPa, that is, $L_y(\sigma_{tr\_unloading} = 21$ MPa$)/L_d = 1.0$; the interface new-slip length approaches to the interface debonding length at the reloading stress $\sigma_{tr\_reloading} = 379$ MPa, that is, $L_z(\sigma_{tr\_reloading} = 379$ MPa$)/L_c = 1.0$; and the hysteresis loops correspond to the interface slip Case 1. Under $\sigma_{max} = 450$ MPa, the interface debonding length occupies 53% of the matrix crack spacing, that is, $2L_d(\sigma_{max} = 450$ MPa$)/L_c = 53\%$; the interface counter-slip length approaches to 94% of the interface debonding length, that is, $L_y(\sigma_{min})/L_d = 94\%$; the interface new-slip length approaches to 94% of the interface debonding length, that is, $L_z(\sigma_{max})/L_d = 94\%$; and the hysteresis loops correspond to the interface slip Case 2.

2.4.3 Effect of Interface Debonding Energy on Mechanical Hysteresis Loops

The hysteresis loops and interface slip of mini-SiC/SiC composite under σ_{max} = 350, 400, and 450 MPa for Γ_i = 4 and 5 J/m^2 are shown in Figure 2.9. When the interface debonding energy increases, the interface debonding and slip range decrease, and the hysteresis loops area and peak strain decrease, and the hysteresis modulus increases.

When Γ_i = 4.0 J/m^2 at σ_{max} = 350 MPa, the interface debonding length occupies 46.4% of the matrix crack spacing, that is, $2L_d(\sigma_{max}$ = 350 MPa)/L_c = 46.4%; the interface counter-slip length approaches to the interface debonding length at the unloading stress $\sigma_{tr\_unloading}$ = 100 MPa, that is, $L_y(\sigma_{tr\_unloading}$ = 100 MPa)/L_d = 1; the interface new-slip length approaches to the interface debonding length at the reloading stress $\sigma_{tr\_reloading}$ = 250 MPa, that is, $L_z(\sigma_{tr\_reloading}$ = 250 MPa)/L_d = 1.0; and the

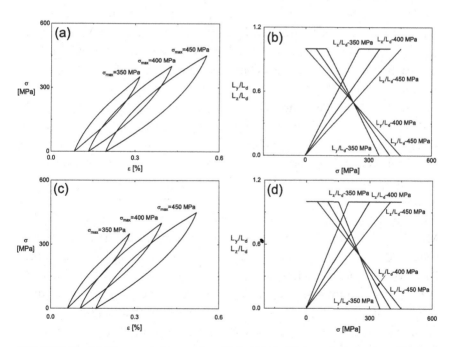

FIGURE 2.9 (a) Hysteresis loops when Γ_i = 4 J/m^2, (b) interface slip length when Γ_i = 4 J/m^2, (c) hysteresis loops when Γ_i = 5 J/m^2, and (d) interface slip length when Γ_i = 5 J/m^2.

hysteresis loops correspond to the interface slip Case 1. Under $\sigma_{max} = 400$ MPa, the interface debonding length occupies 65% of the matrix crack spacing, that is, $2L_d(\sigma_{max} = 400$ MPa$)/L_c = 65\%$; the interface counter-slip length approaches to the interface debonding length at the unloading stress $\sigma_{tr\_unloading} = 50$ MPa, that is, $L_y(\sigma_{tr\_unloading} = 50$ MPa$)/L_d = 1.0$; the interface new-slip length approaches to the interface debonding length at the reloading stress $\sigma_{tr\_reloading} = 350$ MPa, that is, $L_z(\sigma_{tr\_reloading} = 350$ MPa$)/L_c = 1.0$; and the hysteresis loops correspond to the interface slip Case 1. Under $\sigma_{max} = 450$ MPa, the interface debonding length occupies 83.5% of the matrix crack spacing, that is, $2L_d(\sigma_{max} = 450$ MPa$)/L_c = 83.5\%$; the interface counter-slip length approaches to 99% of the interface debonding length, that is, $L_y(\sigma_{min})/L_d = 99\%$; the interface new-slip length approaches to 99% of the interface debonding length, that is, $L_z(\sigma_{max})/L_d = 99\%$; and the hysteresis loops correspond to the interface slip Case 2.

When $\Gamma_i = 5$ J/m^2 at $\sigma_{max} = 350$ MPa, the interface debonding length occupies 36.8% of the matrix crack spacing, that is, $2L_d(\sigma_{max} = 350$ MPa$)/L_c = 36.8\%$; the interface counter-slip length approaches to the interface debonding length at the unloading stress $\sigma_{tr\_unloading} = 152$ MPa, that is, $L_y(\sigma_{tr\_unloading} = 152$ MPa$)/L_d = 1.0$ and the interface new-slip length approaches to the interface debonding length at the reloading stress $\sigma_{tr\_reloading} = 198$ MPa, that is, $L_z(\sigma_{tr\_reloading} = 198$ MPa$)/L_d = 1.0$; and the hysteresis loops correspond to the interface slip Case 1. Under $\sigma_{max} = 400$ MPa, the interface debonding length occupies 55.4% of the matrix crack spacing, that is, $2L_d(\sigma_{max} = 400$ MPa$)/L_c = 55.4\%$; the interface counter-slip length approaches to the interface debonding length at the unloading stress $\sigma_{tr\_unloading} = 102$ MPa, that is, $L_y(\sigma_{tr\_unloading} = 102$ MPa$)/L_d = 1.0$; the interface new-slip length approaches to the interface debonding length at the reloading stress $\sigma_{tr\_reloading} = 298$ MPa, that is, $L_z(\sigma_{tr\_reloading} = 298$ MPa$)/L_c = 1.0$; and the hysteresis loops correspond to the interface slip Case 1. Under $\sigma_{max} = 450$ MPa, the interface debonding length occupies 73.9% of the matrix crack spacing, that is, $2L_d(\sigma_{max} = 450$ MPa$)/L_c = 73.9\%$; the interface counter-slip length approaches to the interface debonding length at the unloading stress $\sigma_{tr\_unloading} = 52$ MPa, that is, $L_y(\sigma_{tr\_unloading} = 52$ MPa$)/L_d = 1.0$; the interface new-slip length approaches to the interface debonding length at the reloading stress $\sigma_{tr\_reloading} = 398$ MPa, that is, $L_z(\sigma_{tr\_reloading} = 398$ MPa$)/L_c = 1.0$; and the hysteresis loops correspond to the interface slip Case 1.

2.4.4 Effect of Matrix Cracking on Mechanical Hysteresis Loops

The hysteresis loops and interface slip of mini-SiC/SiC composite under σ_{max} = 350, 400, and 450 MPa for L_c = 250 and 300 μm are shown in Figure 2.10. When the matrix cracking space increases, the interface debonding and slip range decrease, the hysteresis loops area and peak strain decrease, and the hysteresis modulus increases.

When L_c = 250 μm at σ_{max} = 350 MPa, the interface debonding length occupies 42.2% of the matrix crack spacing, that is, $2L_d(\sigma_{max}=350\,MPa)/L_c$ = 42.2%; the interface counter-slip length approaches to the interface debonding length at the unloading stress $\sigma_{tr\_unloading}$ = 65 MPa, that is, $2L_y(\sigma_{tr\_unloading}$ = 65 MPa)$/L_c$ = 42.2%; the interface new-slip length approaches to the interface debonding length at the reloading stress $\sigma_{tr\_reloading}$ = 285 MPa, that is, $2L_z(\sigma_{tr\_reloading}$ = 285 MPa)$/L_c$ = 42.2%; and the hysteresis loops correspond to the interface slip Case 1. Under σ_{max} = 400 MPa, the interface debonding length occupies 57.1% of the matrix crack

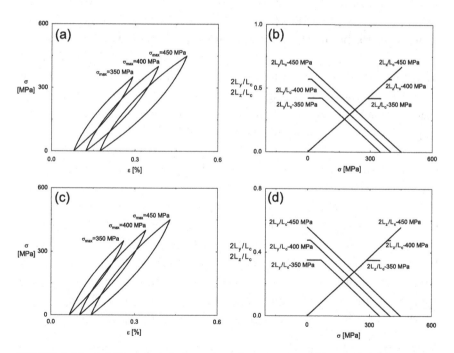

FIGURE 2.10   (a) Hysteresis loops when L_c = 250 μm, (b) interface slip length when L_c = 250 μm, (c) hysteresis loops when L_c = 300 μm, and (d) interface slip length when L_c = 300 μm.

spacing, that is, $2L_d(\sigma_{max} = 400 \text{ MPa})/L_c = 57.1\%$; the interface counter-slip length approaches to the interface debonding length at the unloading stress $\sigma_{tr\_unloading} = 15 \text{ MPa}$, that is, $2L_y(\sigma_{tr\_unloading} = 15 \text{ MPa})/L_c = 57.1\%$; the interface new-slip length approaches to the interface debonding length at the reloading stress $\sigma_{tr\_reloading} = 385 \text{ MPa}$, that is, $2L_z(\sigma_{tr\_reloading} = 385 \text{ MPa})/L_c = 57.1\%$; and the hysteresis loops correspond to the interface slip Case 1. Under $\sigma_{max} = 450 \text{ MPa}$, the interface debonding length occupies 71.9% of the matrix crack spacing, that is, $2L_d(\sigma_{max} = 450 \text{ MPa})/L_c = 71.9\%$; the interface counter-slip length approaches to 66.8% of the matrix crack spacing, that is, $2L_y(\sigma_{min})/L_c = 66.8\%$; the interface new-slip length approaches to 66.8% of the matrix crack spacing, that is, $2L_z(\sigma_{max})/L_c = 66.8\%$; and the hysteresis loops correspond to the interface slip Case 2.

When $L_c = 300 \text{ μm}$ at $\sigma_{max} = 350 \text{ MPa}$, the interface debonding length occupies 35.2% of the matrix crack spacing, that is, $2L_d(\sigma_{max} = 350 \text{ MPa})/L_c = 35.2\%$; the interface counter-slip length approaches to the interface debonding length at the unloading stress $\sigma_{tr\_unloading} = 65 \text{ MPa}$, that is, $2L_y(\sigma_{tr\_unloading} = 65 \text{ MPa})/L_c = 35.2\%$; the interface new-slip length approaches to the interface debonding length at the reloading stress $\sigma_{tr\_reloading} = 285 \text{ MPa}$, that is, $2L_z(\sigma_{tr\_reloading} = 285 \text{ MPa})/L_c = 35.2\%$; and the hysteresis loops correspond to the interface slip Case 1. Under $\sigma_{max} = 400$ MPa, the interface debonding length occupies 47.6% of the matrix crack spacing, that is, $2L_d(\sigma_{max} = 400 \text{ MPa})/L_c = 47.6\%$; the interface counter-slip length approaches to the interface debonding length at the unloading stress $\sigma_{tr\_unloading} = 15 \text{ MPa}$, that is, $2L_y(\sigma_{tr\_unloading} = 15 \text{ MPa})/L_c = 47.6\%$; the interface new-slip length approaches to the interface debonding length at the reloading stress $\sigma_{tr\_reloading} = 385 \text{ MPa}$, that is, $2L_z(\sigma_{tr\_reloading} = 385 \text{ MPa})/L_c = 47.6\%$; and the hysteresis loops correspond to the interface slip Case 1. Under $\sigma_{max} = 450 \text{ MPa}$, the interface debonding length occupies 59.9% of the matrix crack spacing, that is, $2L_d(\sigma_{max} = 450 \text{ MPa})/L_c = 59.9\%$; the interface counter-slip length approaches to 55.6% of the matrix crack spacing, that is, $2L_y(\sigma_{min})/L_c = 55.6\%$; the interface new-slip length approaches to 55.6% of the matrix crack spacing, that is, $2L_z(\sigma_{max})/L_c = 55.6\%$; and the hysteresis loops correspond to the interface slip Case 2.

2.4.5 Effect of Fiber Failure on Mechanical Hysteresis Loops

The broken fiber fraction, hysteresis loops, and interface slip of mini-SiC/SiC composite under $\sigma_{max} = 400 \text{ MPa}$ for the fiber strength $\sigma_0 = 3$ and 3.6 GPa are shown in Figure 2.11. When the fiber strength increases, the

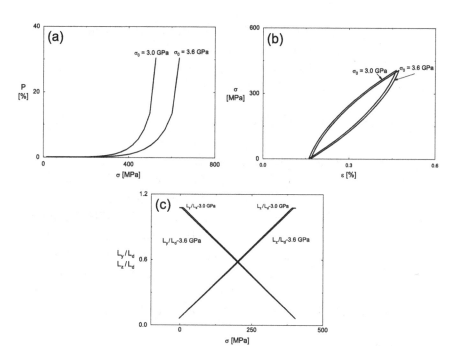

FIGURE 2.11   (a) Broken fibers fraction; (b) Hysteresis loops; and (c) Interface slip length when σ_0 = 3.0 and 3.6 GPa.

interface debonding and slip range decrease, the hysteresis loops area and peak strain decrease, and the hysteresis modulus increases.

When σ_0 = 3 GPa, the tensile strength of mini-SiC/SiC composite is σ_{uts} = 531 MPa, and the broken fiber volume fraction is P = 2.5%, the interface debonding length occupies 73.7% of the matrix crack spacing, that is, $2L_d(\sigma_{max}$ = 400 MPa)$/L_c$ = 73.7%; the interface counter-slip length approaches to the interface debonding length at the unloading stress $\sigma_{tr\_unloading}$ = 12 MPa, that is, $L_y(\sigma_{tr\_unloading}$ = 12 MPa)$/L_d$ = 1.0; the interface new-slip length approaches to the interface debonding length at the reloading stress $\sigma_{tr\_reloading}$ = 388 MPa, that is, $L_z(\sigma_{tr\_reloading}$ = 388 MPa)$/L_d$ = 1.0; and the hysteresis loops correspond to the interface slip Case 1.

When σ_0 = 3.6 GPa, the tensile strength of mini-SiC/SiC composite is σ_{uts} = 637 MPa, and the broken fiber volume fraction is P = 0.8%, the interface debonding length occupies 72% of the matrix crack spacing, that is, $2L_d(\sigma_{max}$ = 400 MPa)$/L_c$ = 72%; the interface counter-slip length approaches to the interface debonding length at the unloading stress $\sigma_{tr\_unloading}$ = 4 MPa, that is, $L_y(\sigma_{tr\_unloading} = 4\ MPa)/L_d = 1.0$; the interface new-slip length approaches to the interface debonding length at the reloading stress $\sigma_{tr\_reloading} = 396\ MPa$, that is, $L_z(\sigma_{tr\_reloading} = 396\ MPa)/L_d = 1.0$; and the hysteresis loops correspond to the interface slip Case 1.

2.5 SUMMARY AND CONCLUSION

In this chapter, the mechanical hysteresis behavior of SiC/SiC minicomposites was investigated. The effects of constituent properties and damage state on the mechanical hysteresis loops and interface slip of SiC/SiC minicomposite were analyzed. Experimental matrix cracking density, mechanical hysteresis loops, and the interface slip of Hi-Nicalon™, Hi-Nicalon™ Type S, and Tyranno™ ZMI SiC/SiC minicomposites are predicted. The interface slip range decreases with increasing fiber volume, interface shear stress, interface debonding energy, and increases with fiber fracture. The stress-dependent mechanical hysteresis of mini-CMCs can be used to monitor the damage evolution of woven CMCs.

REFERENCES

1. Cho CD, Holmes JW, Barber JR. Estimation of interfacial shear in ceramic composites from frictional heating measurements. *J. Am. Ceram. Soc.* 1991; 74:2802–2808.
2. Li LB. Relationship between hysteresis dissipated energy and temperature rising in fiber-reinforced ceramic-matrix composites under cyclic loading. *Appl. Compos. Mater.* 2016; 23:337–355.
3. Fantozzi G, Reynaud P. Mechanical hysteresis in ceramic matrix composites. *Mater. Sci. Eng. A* 2009; 521–522:18–23.
4. Mei H, Cheng L. Comparisons of the mechanical hysteresis of carbon/ceramic-matrix composites with different fiber preforms. *Carbon* 2009; 47:1034–1042.
5. Li LB. Hysteresis loops of carbon fiber-reinforced ceramic-matrix composites with different fiber preforms. *Ceram. Int.* 2016; 42:16535–16551.
6. Morscher GN, Martinez-Fernandez, J. Determination of interfacial properties using a single-fiber microcomposite test. *J. Am. Ceram. Soc.* 1996; 79:1083–1091.
7. Morscher GN. Tensile stress rupture of SiC/SiC minicomposites with carbon and boron nitride interphases at elevated temperatures in air. *J. Am. Ceram. Soc.* 2007; 80:2029–2042.
8. Bertrand S, Forio P, Pailler R, Lamon J. Hi-Nicalon/SiC minicomposites with (Pyrocarbon/SiC)$_n$ nanoscale multilayered interphases. *J. Am. Ceram. Soc.* 1999; 82:2464–2473.

9. Martinez-Fernandez J, Morscher GN. Room and elevated temperature tensile properties of single tow Hi-Nicalon, carbon interphase, CVI SiC matrix minicomposites. *J. Euro. Ceram. Soc.* 2000; 20:2627–2636.

10. Li LB. Synergistic effects of interface slip and fiber fracture on stress-dependent mechanical hysteresis of SiC/SiC minicomposites. *Compos Interfaces* 2020; 27:937–951.

11. Maillet E, Godin N, R'Mili M, Reynaud P, Fantozzi G, Lamon J. Damage monitoring and identification in SiC/SiC minicomposites using combined acousto-ultrasonics and acoustic emission. *Compos. Part A* 2014; 57:8–15.

12. Almansour A, Maillet E, Ramasamy S, Morscher GN. Effect of fiber content on single tow SiC minicomposite mechanical and damage properties using acoustic emission. *J. Euro. Ceram. Soc.* 2015; 35:3389–3399.

13. Zhang S, Gao X, Chen J, Dong H, Song Y. Strength model of the matrix element in SiC/SiC composites. *Mater. Design* 2016; 101:66–71.

14. Kotil T, Holmes JW, Comninou M. Origin of hysteresis observed during fatigue of ceramic matrix composites. *J. Am. Ceram. Soc.* 1990; 73:1879–1883.

15. Pryce AW, Smith PA. Matrix cracking in unidirectional ceramic matrix composites under quasi-static and cyclic loading. *Acta Metall. Mater.* 1993; 41:1269–1281.

16. Evans AG, Zok FW, McMeeking RM. Fatigue of ceramic matrix composites. *Acta Metall. Mater.* 1995; 43:859–875.

17. Keith WP, Kedward KT. The stress-strain behavior of a porous unidirectional ceramic matrix composites. *Compos.* 1995; 26:163–174.

18. Solti JP, Mall S, Robertson DD. Modeling damage in unidirectional ceramic-matrix composites. *Compos. Sci. Technol.* 1995; 54:55–66.

19. Reynaud P. Cyclic fatigue of ceramic-matrix composites at ambient and elevated temperatures. *Compos. Sci. Technol.* 1996; 56:809–814.

20. Ahn BK, Curtin WA. Strain and hysteresis by stochastic matrix cracking in ceramic matrix composites. *J. Mech. Phys. Solids.* 1997; 45:177–209.

21. Li LB. Fatigue hysteresis behavior of cross-ply C/SiC ceramic matrix composites at room and elevated temperatures. *Mater. Sci. Eng. A* 2013; 586:160–170.

High-Temperature Cyclic-Fatigue Mechanical Hysteresis Behavior in Two-Dimensional Plain-Woven Chemical Vapor Infiltration SiC/SiC Composites

3.1 INTRODUCTION

Under cyclic-fatigue loading of two-dimensional (2D) plain-woven ceramic-matrix composites (CMCs), matrix cracking and fiber/matrix interface debonding occur throughout the CMCs, resulting in behavior that gives rise to a curve exhibiting a hysteresis loop [1, 2]. The hysteresis loops appear as a result of fiber slips relative to the matrix in the interface debonded region [3]. The shape, location, and area of hysteresis loops can reveal the internal damage evolution of CMCs subjected to cyclic loading [4]. Michael [5] investigated the tension-tension cyclic-fatigue behavior of 2D plain-woven SiC/SiC composite at 1000°C in air and steam. It was found that ratcheting, defined as a progressive accumulation of strain with

DOI: 10.1201/b23026-3

an increasing number of cycles, continues throughout the test. Jacob [6] investigated the tension-tension cyclic-fatigue behavior of 2D plain-woven SiC/SiC composite at 1200°C in air and steam. The residual strain, shape, and area of the hysteresis loops change continually throughout cycling. Zhu et al. [7] investigated the tension-tension cyclic-fatigue behavior of an enhanced SiC/SiC composite at a temperature of 1300°C in air condition and analyzed the evolution of the stress-strain hysteresis loops. The slope decreases and the width of the hysteresis loops increases as the number of cycles increases. The former indicates a decrease of the modulus, and the latter means a decrease in the interfacial sliding resistance. Kuo and Chou [8] investigated matrix multi-cracking in cross-ply CMCs and classified the multiple cracking states into five modes, in which cracking mode 3 and mode 5 involve matrix cracking and interface debonding in the longitudinal plies. It was found that the increase in the transverse ply thickness reduces the matrix cracking stress in the longitudinal plies.

In this chapter, the high-temperature cyclic-fatigue mechanical hysteresis behavior of 2D plain-woven SiC/SiC composites is investigated. The interface slip between fibers and the matrix existed in matrix cracking mode 3 and mode 5, in which matrix cracking and interface debonding occurred in the longitudinal yarns, are considered to be the major reason for hysteresis loops of 2D plain-woven CMCs. The hysteresis loops of 2D plain-woven SiC/SiC composite corresponding to different peak stresses, test conditions, and loading frequencies are predicted.

3.2 MICROMECHANICAL HYSTERESIS CONSTITUTIVE MODEL

Under cyclic-fatigue loading, Figure 3.1 shows the five different types of the matrix cracking modes in 2D plain-woven CMCs [9]:

- Mode 1, transverse cracking in transverse yarns

- Mode 2: transverse cracking and matrix cracking occurred in transverse and longitudinal yarns with perfect fiber/matrix interface bonding in longitudinal yarns

- Mode 3: transverse cracking and matrix cracking occurred in transverse and longitudinal yarns with interface debonding in longitudinal yarns

- Mode 4: matrix cracking in longitudinal yarns with interface bonding

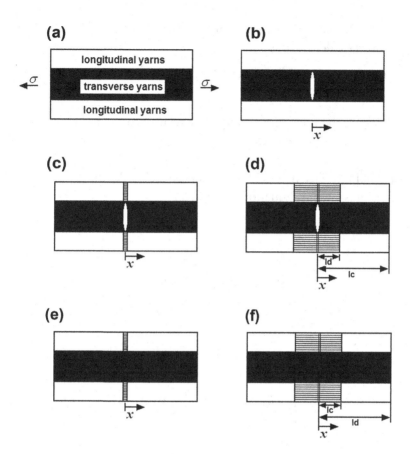

FIGURE 3.1 (a) Undamaged composite, (b) mode 1, (c) mode 2, (d) mode 3, (e) mode 4, and (f) mode 5 of 2D plain-woven CMCs.

- Mode 5: matrix cracking in longitudinal yarns with interface debonding

On unloading and subsequent reloading, the frictional slip occurred between fibers and the matrix in the longitudinal yarns is the major reason for hysteresis loops of 2D plain-woven CMCs.

For matrix cracking mode 3, the hysteresis loops can be divided into four different cases:

- Case 1, the fiber slides completely relative to the matrix with complete interface debonding

- Case 2, the fiber slides partially relative to the matrix with partially interface debonding

- Case 3, the fiber slides partially relative to the matrix with complete interface debonding

- Case 4, the fiber slides completely relative to the matrix with complete interface debonding

For partial interface debonding, the composite's unloading and reloading strain is presented in Equations 3.1 and 3.2:

$$\varepsilon_{unloading} = \frac{\sigma}{V_{f\_axial}E_f} + 4\frac{\tau_i}{E_f}\frac{L_y^2}{r_fL_c}$$
$$-2\frac{\tau_i}{E_f}\frac{(2L_y-l_d)(2L_y-L_c+L_d)}{r_fL_c} - (\alpha_c - \alpha_f)\Delta T, \tag{3.1}$$

$$\varepsilon_{reloading} = \frac{\sigma}{V_{f\_axial}E_f} - 4\frac{\tau_i}{E_f}\frac{L_z^2}{r_fL_c} + \frac{4\tau_i}{E_f}\frac{(L_y-2L_z)^2}{r_fL_c}$$
$$+2\frac{\tau_i}{E_f}\frac{(L_d-2L_y+2L_z)(L_d+2L_y-2L_z-L_c)}{r_fL_c} \tag{3.2}$$
$$-(\alpha_c - \alpha_f)\Delta T,$$

where V_{f_axial} denotes the fiber volume content in the longitudinal direction; E_f denotes the fiber elastic modulus; r_f denotes the fiber radius; τ_i denotes the fiber/matrix interface shear stress in longitudinal yarns; L_c denotes the matrix crack spacing; L_d denotes the interface debonded length; α_f and α_c denote the fiber's and composite's thermal expansion coefficient, respectively; ΔT denotes the temperature difference between fabricated temperature T_0 and room temperature T_1 ($\Delta T = T_1 - T_0$); and L_y and L_z denote the interface counter-slip length and interface new-slip length, respectively.

For complete interface debonding, the composite's unloading and reloading strain is presented in Equations 3.3 and 3.4.

$$\varepsilon_{\text{unloading}} = \frac{\sigma}{V_{\text{f\_axial}}E_{\text{f}}} + 4\frac{\tau_{\text{i}}}{E_{\text{f}}}\frac{L_y^2}{r_{\text{f}}L_{\text{c}}}$$

$$-2\frac{\tau_{\text{i}}}{E_{\text{f}}}\frac{\left(2L_y - L_{\text{c}}/2\right)^2}{r_{\text{f}}L_{\text{c}}} - \left(\alpha_{\text{c}} - \alpha_{\text{f}}\right)\Delta T \tag{3.3}$$

$$\varepsilon_{\text{reloading}} = \frac{\sigma}{V_{\text{f\_axial}}E_{\text{f}}} - 4\frac{\tau_{\text{i}}}{E_{\text{f}}}\frac{L_z^2}{r_{\text{f}}L_{\text{c}}} + 4\frac{\tau_{\text{i}}}{E_{\text{f}}}\frac{\left(L_y - 2L_z\right)^2}{r_{\text{f}}L_{\text{c}}}$$

$$-2\frac{\tau_{\text{i}}}{E_{\text{f}}}\frac{\left(L_{\text{c}}/2 - 2L_y + 2L_z\right)^2}{r_{\text{f}}L_{\text{c}}} - \left(\alpha_{\text{c}} - \alpha_{\text{f}}\right)\Delta T \tag{3.4}$$

For matrix cracking mode 5, the hysteresis loops can also be divided into four different cases. For partial interface debonding, the composite's unloading and reloading strain is presented in Equations 3.5 and 3.6:

$$\varepsilon_{\text{unloading}} = \frac{1}{V_{\text{f\_axial}}E_{\text{f}}}\left(\sigma - k\sigma_{\text{to}}\right) + 4\frac{\tau_{\text{i}}}{E_{\text{f}}}\frac{L_y^2}{r_{\text{f}}L_{\text{c}}}$$

$$-2\frac{\tau_{\text{i}}}{E_{\text{f}}}\frac{\left(2L_y - L_{\text{d}}\right)\left(2L_y + L_{\text{d}} - L_{\text{c}}\right)}{r_{\text{f}}L_{\text{c}}} - \left(\alpha_{\text{c}} - \alpha_{\text{f}}\right)\Delta T, \tag{3.5}$$

$$\varepsilon_{\text{reloading}} = \frac{1}{V_{\text{f\_axial}}E_{\text{f}}}\left(\sigma - k\sigma_{\text{to}}\right) - 4\frac{\tau_{\text{i}}}{E_{\text{f}}}\frac{L_z^2}{r_{\text{f}}L_{\text{c}}} + \frac{4\tau_{\text{i}}}{E_{\text{f}}}\frac{\left(L_y - 2L_z\right)^2}{r_{\text{f}}L_{\text{c}}}$$

$$+2\frac{\tau_{\text{i}}}{E_{\text{f}}}\frac{\left(L_{\text{d}} - 2L_y + 2L_z\right)\left(L_{\text{d}} + 2L_y - 2L_z - L_{\text{c}}\right)}{r_{\text{f}}L_{\text{c}}} \tag{3.6}$$

$$-\left(\alpha_{\text{c}} - \alpha_{\text{f}}\right)\Delta T,$$

where k denotes the proportion of transverse yarns in the entire composite.

For complete interface debonding, the composite's unloading and reloading strain is presented in Equations 3.7 and 3.8.

$$\varepsilon_{\text{unloading}} = \frac{1}{V_{\text{f\_axial}}E_{\text{f}}}\left(\sigma - k\sigma_{\text{to}}\right) + 4\frac{\tau_{\text{i}}}{E_{\text{f}}}\frac{L_y^2}{r_{\text{f}}L_{\text{c}}}$$

$$-2\frac{\tau_{\text{i}}}{E_{\text{f}}}\frac{\left(2L_y - L_{\text{c}}/2\right)^2}{r_{\text{f}}L_{\text{c}}} - \left(\alpha_{\text{c}} - \alpha_{\text{f}}\right)\Delta T \tag{3.7}$$

$$\varepsilon_{\text{reloading}} = \frac{1}{V_{f\_\text{axial}}E_f}\left(\sigma - k\sigma_{\text{to}}\right) - 4\frac{\tau_i}{E_f}\frac{L_z^2}{r_f L_c} + 4\frac{\tau_i}{E_f}\frac{\left(L_y - 2L_z\right)^2}{r_f L_c}$$

$$-2\frac{\tau_i}{E_f}\frac{\left(L_c/2 - 2L_y + 2L_z\right)^2}{r_f L_c} - \left(\alpha_c - \alpha_f\right)\Delta T$$

(3.8)

Considering the effect of multiple matrix cracking modes on cyclic-fatigue hysteresis loops of 2D plain-woven CMCs, the unloading and reloading strains of the composite are presented in Equations 3.9 and 3.10:

$$\left(\varepsilon_{\text{unloading}}\right)_c = \psi\left(\varepsilon_{\text{unloading}}\right)_3 + \left(1 - \psi\right)\left(\varepsilon_{\text{unloading}}\right)_5,$$

(3.9)

$$\left(\varepsilon_{\text{reloading}}\right)_c = \eta\left(\varepsilon_{\text{reloading}}\right)_3 + \left(1 - \eta\right)\left(\varepsilon_{\text{reloading}}\right)_5,$$

(3.10)

where $(\varepsilon_{\text{unloading}})_c$ and $(\varepsilon_{\text{reloading}})_c$ denote the unloading and reloading strain of the composite, respectively; $(\varepsilon_{\text{unloading}})_3$ and $(\varepsilon_{\text{reloading}})_3$ denote the unloading and reloading strain of the matrix cracking mode 3, respectively; $(\varepsilon_{\text{unloading}})_5$ and $(\varepsilon_{\text{reloading}})_5$ denote the unloading and reloading strain of the matrix cracking mode 5, respectively; and ψ is the damage parameter determined by the composite's damage condition, that is, the proportion of matrix cracking mode 3 in the entire of matrix cracking modes of the composite, $\psi \in [0,1]$.

3.3 EXPERIMENTAL COMPARISONS

The tension-tension cyclic-fatigue behavior of 2D plain-woven SiC/SiC composite at $T = 1000$, 1200, and 1300 °C in air and steam conditions were investigated. The materials analyzed are different, based on their fiber volume fraction, and test conditions, that is, testing temperature and testing parameters (loading frequency f, and stress ratio R), are different for these materials. Material properties and testing parameters are given in Table 3.1. The cyclic-fatigue hysteresis loops of 2D plain-woven SiC/SiC composite under different fatigue peak stresses and applied cycle numbers are predicted.

3.3.1 Cyclic-Fatigue Hysteresis Loops at 1000°C in Air

Michael [5] investigated the tension-tension cyclic-fatigue behavior of 2D plain-woven SiC/SiC composite at $T = 1000$ °C in an air condition. The

TABLE 3.1   Material Properties and Testing Parameters of 2D Plain-Woven SiC/SiC Composites

Materials	f	R	V_f /(%)	E_f /GPa	E_m /GPa	r_f /μm	Γ_i /(J/m²)	α_f $(10^{-6}/°C)$	α_m $(10^{-6}/°C)$	ΔT /(°C)
2D-SiC/SiC [5]	1.0	0.1	21.5	150	60	7.5	0.1	4.6	4.38	−400
2D-SiC/SiC [6]	0.1/1.0	0.05	34.8	150	100	7.5	0.1	4.6	4.38	−200
2D-SiC/SiC [7]	20	0.1	40	190	70	6	0.1	4.6	4.38	−100

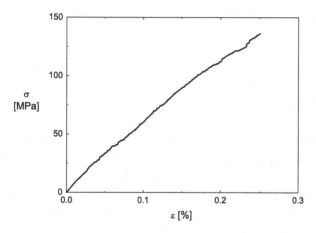

FIGURE 3.2   Monotonic tensile curve of 2D plain-woven SiC/SiC composite at 1000°C in air.

material used for tension-tension fatigue tests underwent a chemical vapor infiltration (CVI) process to produce the interphase and a polymer infiltration and pyrolysis (PIP) process to produce the matrix within the composite. This material is composed of ceramic-grade (CG) NICALON™ fiber tows in an eight-harness satin weave (8HSW), has a fiber volume fraction of V_f = 21.5% and a less than 5% porosity. The monotonic tensile test was conducted under a constant displacement rate of 0.05 mm/s, and the monotonic tensile strength is 136 MPa. The tensile stress-strain curve of 2D woven SiC/SiC composite at T = 1000 °C in air is illustrated in Figure 3.2.

Under σ_{max} = 80 MPa, the experimental and theoretical hysteresis loops, interface slip of matrix cracking modes 3 and 5 corresponding to N = 2, 1000, 10000, 20000, and 30000 are shown in Figures 3.3 through 3.7.

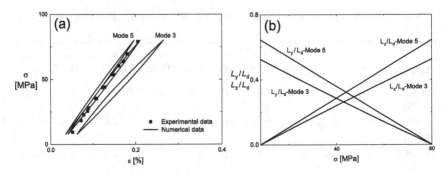

FIGURE 3.3 (a) Experimental and predicted hysteresis loops and (b) the interface slip length of matrix cracking modes 3 and 5 of 2D SiC/SiC composite under σ_{max} = 80 MPa at N = 2.

When N = 2, the fatigue hysteresis loops of matrix cracking modes 3 and 5, the composite experimental data are shown in Figure 3.3a, in which the proportion of matrix cracking mode 3 is ψ = 0.2:

- For matrix cracking mode 3, the hysteresis loops correspond to the interface slip Case 2, as shown in Figure 3.3b. Upon unloading to the valley stress, the interface counter-slip length approaches 52.5% of the interface debonding length, that is, $L_y(\sigma_{min})/L_d$ = 52.5%, and upon reloading to the peak stress, the interface new-slip length approaches 52.5% of the interface debonding length, that is, $L_z(\sigma_{max})/L_d$ = 52.5%.

- For matrix cracking mode 5, the hysteresis loops correspond to the interface slip Case 2. Upon unloading to the valley stress, the interface counter-slip length approaches 64.5% of the interface debonding length, that is, $L_y(\sigma_{min})/L_d$ = 64.5%, and upon reloading to the peak stress, the interface new-slip length approaches 64.5% of the interface debonding length, that is, $L_z(\sigma_{max})/L_d$ = 64.5%.

When N = 1000, the fatigue hysteresis loops of matrix cracking mode 3 and mode 5, the composite and experimental data are shown in Figure 3.4a, in which the proportion of matrix cracking mode 3 is ψ = 0.2:

- For matrix cracking mode 3, the hysteresis loops correspond to the interface slip Case 2, as shown in Figure 3.4b. Upon unloading to the valley stress, the interface counter-slip length approaches 52.3% of

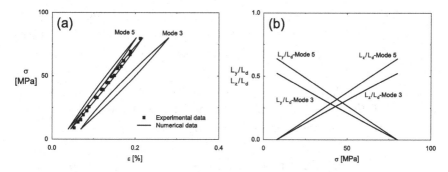

FIGURE 3.4 (a) Experimental and predicted hysteresis loops and (b) the interface slip lengths of matrix cracking modes 3 and 5 of 2D SiC/SiC composite under $\sigma_{\text{max}} = 80$ MPa at $N = 1000$.

the interface debonding length, that is, $L_y(\sigma_{\text{min}})/L_d = 52.3\%$, and upon reloading to the peak stress, the interface new-slip length approaches 52.3% of the interface debonding length, that is, $L_z(\sigma_{\text{max}})/L_d = 52.3\%$.

- For matrix cracking mode 5, the hysteresis loops correspond to the interface slip Case 2. Upon unloading to the valley stress, the interface counter-slip length approaches 63.7% of the interface debonding length, that is, $L_y(\sigma_{\text{min}})/L_d = 63.7\%$, and upon reloading to the peak stress, the interface new-slip length approaches 63.7% of the interface debonding length, that is, $L_z(\sigma_{\text{max}})/L_d = 63.7\%$.

When $N = 10000$, the fatigue hysteresis loops of matrix cracking mode 3 and mode 5, the composite and experimental data are shown in Figure 3.5a, in which the proportion of matrix cracking mode 3 is $\psi = 0.2$:

- For matrix cracking mode 3, the hysteresis loops correspond to the interface slip Case 2, as shown in Figure 3.5b. Upon unloading to the valley stress, the interface counter-slip length approaches 52.2% of the interface debonding length, i.e., $L_y(\sigma_{\text{min}})/L_d = 52.2\%$, and upon reloading to the peak stress, the interface new-slip length approaches 52.2% of the interface debonding length, that is, $L_z(\sigma_{\text{max}})/L_d = 52.2\%$.

- For matrix cracking mode 5, the hysteresis loops correspond to the interface slip Case 2. Upon unloading to the valley stress, the interface counter-slip length approaches 63.4% of the interface debonding length, that is, $L_y(\sigma_{min})/L_d$ = 63.4%, and upon reloading to the peak stress, the interface new-slip length approaches 63.4% of the interface debonding length, that is, $L_z(\sigma_{max})/L_d$ = 63.4%.

When N = 20000, the fatigue hysteresis loops of matrix cracking mode 3 and mode 5, the composite and experimental data are shown in Figure 3.6a, in which the proportion of matrix cracking mode 3 is ψ = 0.2:

- For matrix cracking mode 3, the hysteresis loops correspond to the interface slip Case 2, as shown in Figure 3.6b. Upon unloading to the valley stress, the interface counter-slip length approaches 52.1% of the interface debonding length, that is, $L_y(\sigma_{min})/L_d$ = 52.1%, and upon

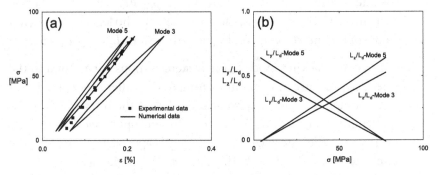

FIGURE 3.5   (a) Experimental and predicted hysteresis loops and (b) the interface slip lengths of matrix cracking modes 3 and 5 of 2D SiC/SiC composite under σ_{max} = 80 MPa at N = 10000.

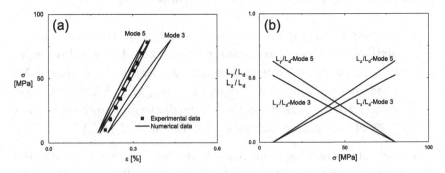

FIGURE 3.6   (a) Experimental and predicted hysteresis loops and (b) the interface slip lengths of matrix cracking modes 3 and 5 of 2D SiC/SiC composite under σ_{max} = 80 MPa at N = 20000.

reloading to the peak stress, the interface new-slip length approaches 52.1% of the interface debonding length, that is, $L_z(\sigma_{max})/L_d = 52.1\%$.

- For matrix cracking mode 5, the hysteresis loops correspond to the interface slip Case 2. Upon unloading to the valley stress, the interface counter-slip length approaches to 63% interface debonding length, that is, $L_y(\sigma_{min})/L_d = 63\%$, and upon reloading to the peak stress, the interface new-slip length approaches 63% of the interface debonding length, that is, $L_z(\sigma_{max})/L_d = 63\%$.

When $N = 30000$, the fatigue hysteresis loops of matrix cracking mode 3 and mode 5, the composite and experimental data are shown in Figure 3.7a, in which the proportion of matrix cracking mode 3 is $\psi = 0.2$:

- For matrix cracking mode 3, the hysteresis loops correspond to the interface slip Case 2, as shown in Figure 3.7b. Upon unloading to the valley stress, the interface counter-slip length approaches 52% of the interface debonding length, that is, $L_y(\sigma_{min})/L_d = 52\%$, and upon reloading to the peak stress, the interface new-slip length approaches 52% of interface debonding length, that is, $L_z(\sigma_{max})/L_d = 52\%$.

- For matrix cracking mode 5, the hysteresis loops correspond to the interface slip Case 2. Upon unloading to the valley stress, the interface counter-slip length approaches 62.7% of the interface debonding length, that is, $L_y(\sigma_{min})/L_d = 62.7\%$, and upon reloading to the

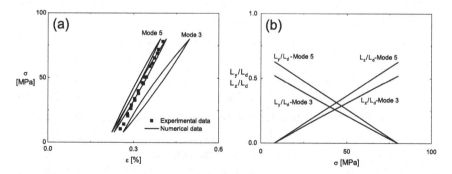

FIGURE 3.7 (a) Experimental and predicted hysteresis loops and (b) the interface slip lengths of matrix cracking modes 3 and 5 of 2D SiC/SiC composite under $\sigma_{max} = 80$ MPa at $N = 30000$.

TABLE 3.2 Damage Parameters and Interface Slip Type of 2D SiC/SiC Composite under σ_{max} = 80 MPa and f = 1 Hz at 1000°C in Air

Items	$N = 2$	$N = 1000$	$N = 10000$	$N = 20000$	$N = 30000$
ψ	0.2	0.2	0.2	0.2	0.2
Cracking mode 3	Case 2	Case 2	Case 2	Case 2	Case 2
Cracking mode 5	Case 2	Case 2	Case 2	Case 2	Case 2

peak stress, the interface new-slip length approaches 62.7% of the interface debonding length, that is, $L_z(\sigma_{max})/L_d$ = 62.7%.

Under fatigue peak stress $\sigma_{max} = 0.58\sigma_{uts}$ at 1000°C in air condition, the proportion of matrix cracking mode 3 occupies 20% of all matrix cracking modes in the 2D SiC/SiC composite; the interface shear stress decreases from τ_i = 15 MPa at N = 2 to τ_i = 10 MPa at N = 30000; and the hysteresis loops of matrix cracking mode 3 and mode 5 both correspond to the interface slip Case 2 from N = 2 to 30000, as shown in Table 3.2.

3.3.2 Cyclic-Fatigue Hysteresis Loops at 1000°C in Steam

Michael [5] investigated the tension-tension cyclic-fatigue behavior of 2D plain-woven SiC/SiC composite at T = 1000°C in a steam condition. The fatigue tests were conducted at the loading frequency f = 1.0 Hz with a stress ratio R = 0.1.

Under σ_{max} = 60 MPa, the experimental and theoretical hysteresis loops, interface slip of matrix cracking modes 3 and 5 corresponding to N = 2, 10000, 100000, 150000, and 190000 are shown in Figures 3.8 through 3.12.

When N = 2, the fatigue hysteresis loops of matrix cracking mode 3 and mode 5, the composite and experimental data are shown in Figure 3.8a, in which the proportion of matrix cracking mode 3 is ψ = 0.2:

- For matrix cracking mode 3, the hysteresis loops correspond to the interface slip Case 2, as shown in Figure 3.8b. Upon unloading to the valley stress, the interface counter-slip length approaches 56.6% of the interface debonding length, i.e., $L_y(\sigma_{min})/L_d$ = 56.6%; and upon reloading to the peak stress, the interface new-slip length approaches 56.6% of the interface debonding length, i.e., $L_z(\sigma_{max})/L_d$ = 56.6%.

- For matrix cracking mode 5, the hysteresis loops correspond to the interface slip Case 2, as shown in Figure 3.8(b). Upon unloading to

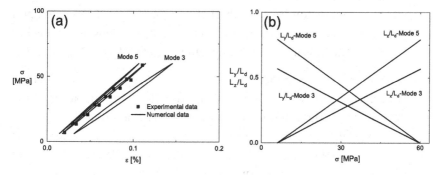

FIGURE 3.8 (a) Experimental and predicted hysteresis loops and (b) the interface slip lengths of matrix cracking modes 3 and 5 of 2D SiC/SiC composite under σ_{max} = 60 MPa at N = 2.

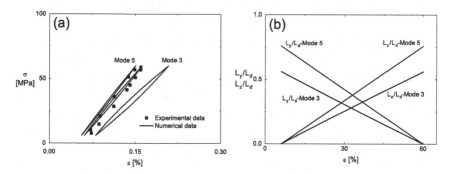

FIGURE 3.9 (a) Experimental and predicted hysteresis loops and (b) the interface slip lengths of matrix cracking modes 3 and 5 of 2D SiC/SiC composite under σ_{max} = 60 MPa at N = 10000.

the valley stress, the interface counter-slip length approaches 79% of the interface debonding length, that is, $L_y(\sigma_{min})/L_d$ = 79%, and upon reloading to the peak stress, the interface new-slip length approaches 79% of the interface debonding length, that is, $L_z(\sigma_{max})/L_d$ = 79%.

When N = 10000, the hysteresis loops of matrix cracking mode 3 and mode 5, the composite and experimental data are shown in Figure 3.9a, in which the proportion of matrix cracking mode 3 is ψ = 0.2:

- For matrix cracking mode 3, the hysteresis loops correspond to the interface slip Case 2, as shown in Figure 3.9b. Upon unloading to the valley stress, the interface counter-slip length approaches 55.8% of

the interface debonding length, that is, $L_y(\sigma_{min})/L_d = 55.8\%$, and upon reloading to the peak stress, the interface new-slip length approaches 55.8% of the interface debonding length, that is, $L_z(\sigma_{max})/L_d = 55.8\%$.

- For matrix cracking mode 5, the hysteresis loops correspond to the interface slip Case 2, as shown in Figure 3.9b. Upon unloading to the valley stress, the interface counter-slip length approaches 75.6% of the interface debonding length, that is, $L_y(\sigma_{min})/L_d = 75.6\%$, and upon reloading to the peak stress, the interface new-slip length approaches 75.6% of the interface debonding length, that is, $L_z(\sigma_{max})/L_d = 75.6\%$.

When $N = 100000$, the fatigue hysteresis loops of matrix cracking mode 3 and mode 5, the composite and experimental data are shown in Figure 3.10a, in which the proportion of matrix cracking mode 3 is $\psi = 0.2$:

- For matrix cracking mode 3, the hysteresis loops correspond to interface slip Case 2, as shown in Figure 3.10b. Upon unloading to the valley stress, the interface counter-slip length approaches 55.5% of the interface debonding length, that is, $L_y(\sigma_{min})/L_d = 55.5\%$; and upon reloading to the peak stress, the interface new-slip length approaches 55.5% of the interface debonding length, that is, $L_z(\sigma_{max})/L_d = 55.5\%$.

- For matrix cracking mode 5, the hysteresis loops correspond to the interface slip Case 2, as shown in Figure 3.10b. Upon unloading to the valley stress, the interface counter-slip length approaches 74.4% of

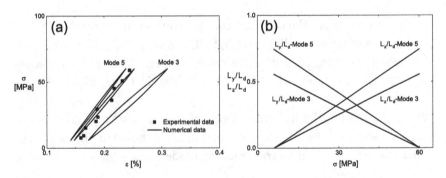

FIGURE 3.10 (a) Experimental and predicted hysteresis loops and (b) the interface slip lengths of matrix cracking modes 3 and 5 of 2D SiC/SiC composite under $\sigma_{max} = 60$ MPa at $N = 100000$.

the interface debonding length, that is, $L_y(\sigma_{min})/L_d = 74.4\%$, and upon reloading to the peak stress, the interface new-slip length approaches 74.4% of the interface debonding length, that is, $L_z(\sigma_{max})/L_d = 74.4\%$.

When N = 150000, the fatigue hysteresis loops of matrix cracking mode 3 and mode 5, the composite and experimental data are shown in Figure 3.11a, in which the proportion of matrix cracking mode 3 is $\psi = 0.2$:

- For matrix cracking mode 3, the hysteresis loops correspond to interface slip Case 2, as shown in Figure 3.11b. Upon unloading to the valley stress, the interface counter-slip length approaches 55.1% of the interface debonding length, that is, $L_y(\sigma_{min})/L_d = 55.1\%$, and upon reloading to the peak stress, the interface new-slip length approaches 55.1% of the interface debonding length, that is, $L_z(\sigma_{max})/L_d = 55.1\%$.

- For matrix cracking mode 5, the hysteresis loops correspond to the interface slip Case 2, as shown in Figure 3.11b. Upon unloading to the valley stress, the interface counter-slip length approaches 72.7% of the interface debonding length, that is, $L_y(\sigma_{min})/L_d = 72.7\%$, and upon reloading to the peak stress, the interface new-slip length approaches 72.7% of the interface debonding length, that is, $L_z(\sigma_{max})/L_d = 72.7\%$.

When N = 190000, the fatigue hysteresis loops of matrix cracking mode 3 and mode 5, the composite and experimental data are shown in Figure 3.12(a), in which the proportion of matrix cracking mode 3 is $\psi = 0.2$:

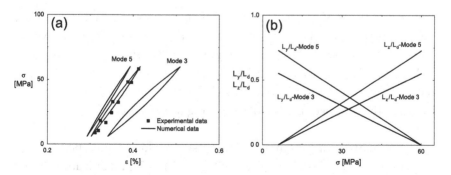

FIGURE 3.11　(a) Experimental and predicted hysteresis loops and (b) the interface slip lengths of matrix cracking modes 3 and 5 of 2D SiC/SiC composite under σ_{max} = 60 MPa at N = 150000.

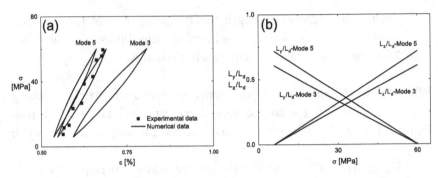

FIGURE 3.12 (a) Experimental and predicted hysteresis loops and (b) the interface slip lengths of matrix cracking mode 3 and mode 5 of 2D SiC/SiC composite under $\sigma_{max} = 60$ MPa at $N = 190000$.

- For matrix cracking mode 3, the hysteresis loops correspond to interface slip Case 3, as shown in Figure 3.12b. Upon unloading to the valley stress, the interface counter-slip length approaches 60.6% of the interface debonding length, that is, $L_y(\sigma_{min})/L_d = 60.6\%$, and upon reloading to the peak stress, the interface new-slip length approaches 60.6% of the interface debonding length, that is, $L_z(\sigma_{max})/L_d = 60.6\%$.

- For matrix cracking mode 5, the hysteresis loops correspond to the interface slip Case 2, as shown in Figure 3.12b. Upon unloading to the valley stress, the interface counter-slip length approaches 71.6% of the interface debonding length, that is, $L_y(\sigma_{min})/L_d = 71.6\%$, and upon reloading to the peak stress, the interface new-slip length approaches 71.6% of the interface debonding length, that is, $L_z(\sigma_{max})/L_d = 71.6\%$.

Under $\sigma_{max} = 100$ MPa, the experimental and theoretical hysteresis loops, interface slip of matrix cracking modes 3 and 5 corresponding to $N = 2$, 500, 3000, and 10000 are shown in Figures 3.13 through 3.16.

When $N = 2$, the fatigue hysteresis loops of matrix cracking mode 3 and mode 5, the composite and experimental data are shown in Figure 3.13a, in which the proportion of matrix cracking mode 3 is $\psi = 0.4$:

- For matrix cracking mode 3, the hysteresis loops correspond to the interface slip Case 2, as shown in Figure 3.13b. Upon unloading to the valley stress, the interface counter-slip length approaches 51.2% of the interface debonding length, that is, $L_y(\sigma_{min})/L_d = 51.2\%$, and upon reloading to the peak stress, the interface new-slip length approaches 51.2% of the interface debonding length, that is, $L_z(\sigma_{max})/L_d = 51.2\%$.

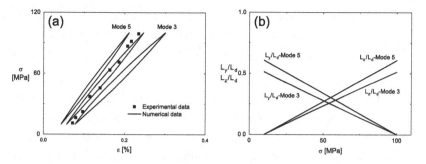

FIGURE 3.13   (a) Experimental and predicted hysteresis loops and (b) the interface slip lengths of matrix cracking modes 3 and 5 of 2D SiC/SiC composite under σ_{max} = 100 MPa at N = 2.

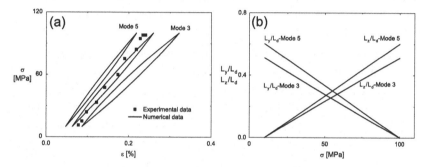

FIGURE 3.14   (a) Experimental and predicted hysteresis loops and (b) the interface slip lengths of matrix cracking modes 3 and 5 of 2D SiC/SiC composite under σ_{max} = 100 MPa at N = 500.

- For matrix cracking mode 5, the hysteresis loops correspond to the interface slip Case 2, as shown in Figure 3.13b. Upon unloading to the valley stress, the interface counter-slip length approaches 60.6% of the interface debonding length, that is, $L_y(\sigma_{min})/L_d$ = 60.6%, and upon reloading to the peak stress, the interface new-slip length approaches 60.6% of the interface debonding length, that is, $L_z(\sigma_{max})/L_d$ = 60.6%.

When N = 500, the fatigue hysteresis loops of matrix cracking mode 3 and mode 5, the composite and experimental data are shown in Figure 3.14a, in which the proportion of matrix cracking mode 3 is ψ = 0.4:

- For matrix cracking mode 3, the hysteresis loops correspond to the interface slip Case 2, as shown in Figure 3.14b. Upon unloading to the valley stress, the interface counter-slip length approaches 51.1% of

the interface debonding length, that is, $L_y(\sigma_{min})/L_d = 51.1\%$, and upon reloading to the peak stress, the interface new-slip length approaches 51.1% of the interface debonding length, that is, $L_z(\sigma_{max})/L_d = 51.1\%$.

- For matrix cracking mode 5, the hysteresis loops correspond to the interface slip Case 2, as shown in Figure 3.14b. Upon unloading to the valley stress, the interface counter-slip length approaches 60.1% of the interface debonding length, that is, $L_y(\sigma_{min})/L_d = 60.1\%$, and upon reloading to the peak stress, the interface new-slip length approaches 60.1% of the interface debonded length, that is, $L_z(\sigma_{max})/L_d = 60.1\%$.

When $N = 3000$, the fatigue hysteresis loops of matrix cracking mode 3 and mode 5, the composite and experimental data are shown in Figure 3.15(a), in which the proportion of matrix cracking mode 3 is $\psi = 0.4$:

- For matrix cracking mode 3, the hysteresis loops correspond to interface slip Case 2, as shown in Figure 3.15b. Upon unloading to the valley stress, the interface counter-slip length approaches 50.8% of the interface debonding length, that is, $L_y(\sigma_{min})/L_d = 50.8\%$, and upon reloading to the peak stress, the interface new-slip length approaches 50.8% of the interface debonding length, that is, $L_z(\sigma_{max})/L_d = 50.8\%$.

- For matrix cracking mode 5, the hysteresis loops correspond to the interface slip Case 2, as shown in Figure 3.15b. Upon unloading to the valley stress, the interface counter-slip length approaches 59.4% of

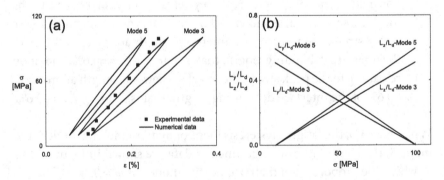

FIGURE 3.15   (a) Experimental and predicted hysteresis loops and (b) the interface slip lengths of matrix cracking modes 3 and 5 of 2D SiC/SiC composite under $\sigma_{max} = 100$ MPa at $N = 3000$.

the interface debonding length, that is, $L_y(\sigma_{min})/L_d = 59.4\%$; and upon reloading to the peak stress, the interface new-slip length approaches 59.4% of the interface debonding length, that is, $L_z(\sigma_{max})/L_d = 59.4\%$.

When $N = 10000$, the fatigue hysteresis loops of matrix cracking mode 3 and mode 5, the composite and experimental data are shown in Figure 3.16a, in which the proportion of matrix cracking mode 3 is $\psi = 0.4$:

• For matrix cracking mode 3, the hysteresis loops correspond to the interface slip Case 2, as shown in Figure 3.16b. Upon unloading to the valley stress, the interface counter-slip length approaches 50.7% of the interface debonding length, that is, $L_y(\sigma_{min})/L_d = 50.7\%$, and upon reloading to the peak stress, the interface new-slip length approaches 50.7% of the interface debonding length, that is, $L_z(\sigma_{max})/L_d = 50.7\%$.

• For matrix cracking mode 5, the hysteresis loops correspond to the interface slip Case 2, as shown in Figure 3.16b. Upon unloading to the valley stress, the interface counter-slip length approaches 58.9% of the interface debonding length, that is, $L_y(\sigma_{min})/L_d = 58.9\%$, and upon reloading to the peak stress, the interface new-slip length approaches 58.9% of the interface debonding length, that is, $L_z(\sigma_{max})/L_d = 58.9\%$.

Under $\sigma_{max} = 60$ MPa at 1000°C in steam, the proportion of matrix cracking mode 3 occupies 20% of all matrix cracking modes in 2D SiC/SiC composite; and the interface shear stress decreases from $\tau_i = 15$ MPa at $N = 2$

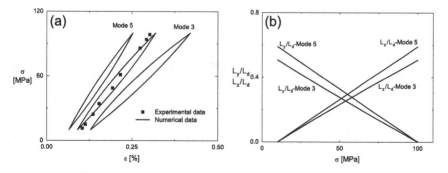

FIGURE 3.16 (a) Experimental and predicted hysteresis loops and (b) the interface slip lengths of matrix cracking modes 3 and 5 of 2D SiC/SiC composite under $\sigma_{max} = 100$ MPa at $N = 10000$.

to τ_i = 3 MPa at N = 190000; and the hysteresis loops of matrix cracking mode 3 and mode 5 correspond to the interface slip Case in 2 and Case 2 at N = 2 to Case 3 and Case 2 at N = 190000. Under σ_{max} = 100 MPa at 1000°C in steam, the proportion of matrix cracking mode 3 occupies 40% of all matrix cracking modes in 2D SiC/SiC composite, and the interface shear stress decreases from τ_i = 15 MPa at N = 2 to τ_i = 8 MPa at N = 10000; and the hysteresis loops of matrix cracking mode 3 and mode 5 both correspond to the interface slip Case 2 from N = 2 to N = 10000.

3.3.3 Cyclic-Fatigue Hysteresis Loops at 1200°C in Air

Jacob [6] investigated the tension-tension cyclic fatigue behavior of 2D plain-woven SiC/SiC composite at T = 1200°C in an air condition. The material used for tension-tension cyclic-fatigue tests was manufactured by CVI of HyprSiC oxidation inhibited matrix material into the woven Hi-Nicalon™ fiber preforms. The composite consisted of eight plies of Hi-Nicalon™ [0°/90°] fabric woven in an 8HSW. Prior to matrix densification, the preforms were coated with pyrolytic carbon with boron carbide overlay to decrease interface bonding between fibers and matrix. The volume fraction of the fibers was V_f = 34.8% and less than 5% porosity. The monotonic tensile test was conducted under a constant displacement rate 0.05 mm/s, and the monotonic tensile strength is 308 MPa. The tensile stress-strain curve of 2D plain-woven SiC/SiC composite at T = 1200 °C in air is shown in Figure 3.17.

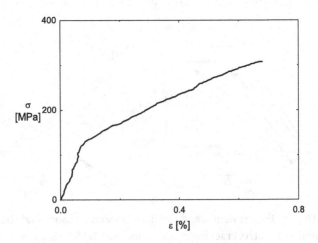

FIGURE 3.17   Monotonic tensile curve of 2D SiC/SiC composite at 1200°C in air.

When σ_{max} = 140 MPa and f = 0.1 Hz, the experimental and theoretical hysteresis loops, interface slip of matrix cracking mode 3 and mode 5 corresponding to N = 1000, 10000, and 30000 are shown in Figures 3.18 through 3.20.

When N = 1000, the fatigue hysteresis loops of matrix cracking mode 3 and mode 5, the composite and experimental data are shown in Figure 3.18a, in which the proportion of matrix cracking mode 3 is ψ = 0.2:

- For matrix cracking mode 3, the hysteresis loops correspond to the interface slip Case 2, as shown in Figure 3.18(b). Upon unloading to the valley stress, the interface counter-slip length approaches 52.9% of the interface debonding length, that is, $L_y(\sigma_{min})/L_d$ = 52.9%, and upon reloading to the peak stress, the interface new-slip length approaches 52.9% of the interface debonding length, that is, $L_z(\sigma_{max})/L_d$ = 52.9%.

- For matrix cracking mode 5, the hysteresis loops correspond to the interface slip Case 2. Upon unloading to the valley stress, the interface counter-slip length approaches 67% of the interface debonding length, i.e., $L_y(\sigma_{min})/L_d$ = 67%, and upon reloading to the peak stress, the interface new-slip length approaches 67% of the interface debonding length, that is, $L_z(\sigma_{max})/L_d$ = 67%.

When N = 10000, the fatigue hysteresis loops of matrix cracking mode 3 and mode 5, the composite and experimental data are shown in Figure 3.19(a), in which the proportion of matrix cracking mode 3 is ψ = 0.2:

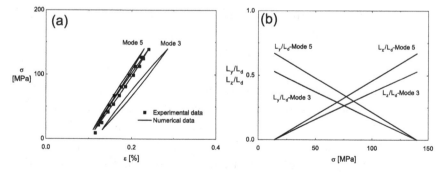

FIGURE 3.18   (a) Experimental and predicted hysteresis loops and (b) the interface slip lengths of matrix cracking modes 3 and 5 of 2D SiC/SiC composite under σ_{max} = 140 MPa at N = 1000.

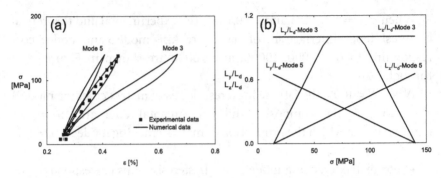

FIGURE 3.19   (a) Experimental and predicted hysteresis loops; and (b) the interface slip lengths of matrix cracking modes 3 and 5 of 2D SiC/SiC composite under σ_{max} = 140 MPa at N = 10000.

- For matrix cracking mode 3, the hysteresis loops correspond to interface slip Case 4, as shown in Figure 3.19b. Upon unloading, the interface counter-slip length approaches the interface debonding length at σ_{tr_fu} = 89.6 MPa, that is, $L_y(\sigma_{tr\_fu})/L_d$ = 1; and upon reloading to σ_{tr_fr} = 64.4 MPa, the interface new-slip length approaches the interface debonding length, that is, $L_z(\sigma_{tr\_fr})/L_d$ = 1.

- For matrix cracking mode 5, the hysteresis loops correspond to the interface slip Case 2. Upon unloading to the valley stress, the interface counter-slip length approaches 64.8% of the interface debonding length, that is, $L_y(\sigma_{min})/L_d$ = 64.8%; and upon reloading to the peak stress, the interface new-slip length approaches 64.8% of the interface debonding length, i.e., $L_z(\sigma_{max})/L_d$ = 64.8%.

When N = 30000, the fatigue hysteresis loops of matrix cracking mode 3 and mode 5, the composite and experimental data are shown in Figure 3.20a, in which the proportion of matrix cracking mode 3 is ψ = 0.2.

- For matrix cracking mode 3, the hysteresis loops correspond to interface slip Case 4, as shown in Figure 3.20b. Upon unloading, the interface counter-slip length approaches the interface debonding length at σ_{tr_fu} = 108.5 MPa, that is, $L_y(\sigma_{tr\_fu})/L_d$ = 1, and upon reloading to σ_{tr_fr} = 45.5 MPa, the interface new-slip length approaches the interface debonding length, that is, $L_z(\sigma_{tr\_fr})/L_d$ = 1.

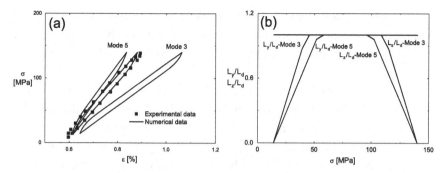

FIGURE 3.20 (a) Experimental and predicted hysteresis loops and (b) the interface slip lengths of matrix cracking modes 3 and 5 of 2D SiC/SiC composite under σ_{max} = 140 MPa at N = 30000.

- For matrix cracking mode 5, the hysteresis loops correspond to the interface slip Case 4. Upon unloading, the interface counter-slip length approaches interface debonding length at σ_{tr_fu} = 95.9 MPa, that is, $L_y(\sigma_{tr\_fu})/L_d$ = 1, and upon reloading to σ_{tr_fr} = 58.1 MPa, the interface new-slip length approaches the interface debonding length, that is, $L_z(\sigma_{tr\_fr})/L_d$ = 1.

When σ_{max} = 140 MPa and f = 1 Hz, the experimental and theoretical predicted hysteresis loops, interface slip of matrix cracking modes 3 and 5 corresponding to N = 10000, 30000, and 60000 are shown in Figures 3.21 through 3.23.

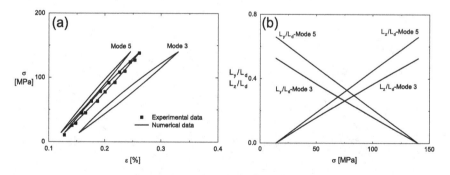

FIGURE 3.21 (a) Experimental and predicted hysteresis loops and (b) the interface slip lengths of matrix cracking modes 3 and 5 of 2D SiC/SiC composite under σ_{max} = 140 MPa at N = 10000.

When $N = 10000$, the fatigue hysteresis loops of matrix cracking modes 3 and 5, the composite and experimental data are shown in Figure 3.21a, in which the proportion of matrix cracking mode 3 is $\psi = 0.2$:

- For matrix cracking mode 3, the hysteresis loops correspond to interface slip Case 2, as shown in Figure 3.21b. Upon unloading to the valley stress, the interface counter-slip length approaches 52.6% of the interface debonding length, that is, $L_y(\sigma_{min})/L_d = 52.6\%$; and upon reloading to the peak stress, the interface new-slip length approaches 52.6% of the interface debonding length, that is, $L_z(\sigma_{max})/L_d = 52.6\%$.

- For matrix cracking mode 5, the hysteresis loops correspond to the interface slip Case 2. Upon unloading to the valley stress, the interface counter-slip length approaches 65.9% of the interface debonding length, that is, $L_y(\sigma_{min})/L_d = 65.9\%$, and upon reloading to the peak stress, the interface new-slip length approaches 65.9% of the interface debonding length, that is, $L_z(\sigma_{max})/L_d = 65.9\%$.

When $N = 30000$, the fatigue hysteresis loops of matrix cracking mode 3 and mode 5, the composite and experimental data are shown in Figure 3.22a, in which the proportion of matrix cracking mode 3 is $\psi = 0.2$:

- For matrix cracking mode 3, the hysteresis loops correspond to interface slip Case 3, as shown in Figure 3.22b. Upon unloading to the valley stress, the interface counter-slip length approaches 98.3% of

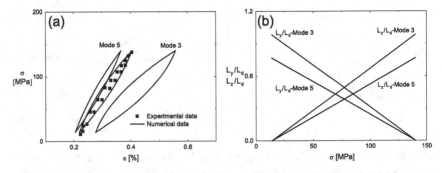

FIGURE 3.22 (a) Experimental and predicted hysteresis loops and (b) the interface slip lengths of matrix cracking mode 3 and mode 5 of 2D SiC/SiC composite under $\sigma_{max} = 140$ MPa at $N = 30000$.

the interface debonding length, that is, $L_y(\sigma_{min})/L_d = 98.3\%$, and upon reloading to the peak stress, the interface new-slip length approaches 98.3% of the interface debonding length, that is, $L_z(\sigma_{max})/L_d = 98.3\%$.

- For matrix cracking mode 5, the hysteresis loops correspond to the interface slip Case 3. Upon unloading to the valley stress, the interface counter-slip length approaches 76.7% of the interface debonding length, that is, $L_y(\sigma_{min})/L_d = 76.7\%$, and upon reloading to the peak stress, the interface new-slip length approaches 76.7% of the interface debonding length, that is, $L_z(\sigma_{max})/L_d = 76.7\%$.

When $N = 60000$, the fatigue hysteresis loops of matrix cracking mode 3 and mode 5, the composite and experimental data are given in Figure 3.23a, in which the proportion of matrix cracking mode 3 is $\psi = 0.2$:

- For matrix cracking mode 3, the hysteresis loops correspond to interface slip Case 4, as shown in Figure 3.23b. Upon unloading, the interface counter-slip length approaches the interface debonding length at $\sigma_{tr\_fu} = 95.9$ MPa, that is, $L_y(\sigma_{tr\_fu})/L_d = 1$, and upon reloading to $\sigma_{tr\_fr} = 58.1$ MPa, the interface new-slip length approaches the interface debonding length, that is, $L_z(\sigma_{tr\_fr})/L_d = 1$.

- For matrix cracking mode 5, the hysteresis loops correspond to the interface slip Case 4. Upon unloading, the interface counter-slip length approaches the interface debonding length at

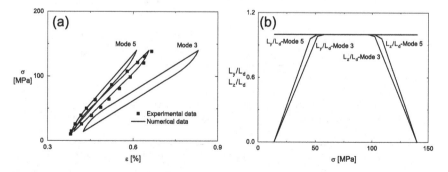

FIGURE 3.23   (a) Experimental and predicted hysteresis loops and (b) the interface slip lengths of matrix cracking modes 3 and 5 of 2D SiC/SiC composite under $\sigma_{max} = 140$ MPa at $N = 60000$.

$\sigma_{tr\_fu} = 102.2$ MPa, that is, $L_y(\sigma_{tr\_fu})/L_d = 1$, and upon reloading to $\sigma_{tr\_fr} = 51.8$, the interface new-slip length approaches the interface debonding length, that is, $L_z(\sigma_{tr\_fr})/L_d = 1$.

Under fatigue peak stress $\sigma_{max} = 0.45\sigma_{uts}$ at $T = 1200°C$ in air, the proportion of matrix cracking mode 3 occupies 20% of all matrix cracking modes in 2D SiC/SiC composite. When the loading frequency is $f = 0.1$ Hz, the interface shear stress decreases from $\tau_i = 15$ MPa at $N = 1000$ to $\tau_i = 2$ MPa at $N = 30000$, and the hysteresis loops of matrix cracking mode 3 and mode 5 correspond to the interface slip Case 2 and Case 2 when $N = 1000$, Case 4 and Case 2 when $N = 10000$, and Case 4 and Case 4 when $N = 30000$, respectively, as shown in Table 3.3. When the loading frequency is 1 Hz, the interface shear stress decreases from $\tau_i = 10$ MPa at $N = 10000$ to $\tau_i = 2$ MPa at $N = 60000$, and the hysteresis loops of matrix cracking mode 3 and mode 5 correspond to interface slip Case 2 and Case 2 when $N = 10000$, Case 3 and Case 3 when $N = 30000$, and Case 4 and Case 4 when $N = 60000$, respectively, as shown in Table 3.4.

3.3.4 Cyclic-Fatigue Hysteresis Loops at 1200°C in Steam

Jacob [6] investigated the tension-tension cyclic-fatigue behavior of 2D plain-woven SiC/SiC composite at $T = 1200$ °C in a steam condition. The

TABLE 3.3 Damage Parameter and Interface Slip Type of 2D SiC/SiC Composite under $\sigma_{max} = 140$ MPa and $f = 0.1$ Hz at $T = 1200$ °C in Air

Items	N = 1000	N = 10000	N = 30000
ψ	0.2	0.2	0.2
Cracking mode 3	Case 2	Case 4	Case 4
Cracking mode 5	Case 2	Case 2	Case 4

TABLE 3.4 Damage Parameter and Interface Slip Type of 2D SiC/SiC Composite under $\sigma_{max} = 140$ MPa and $f = 1$ Hz at $T = 1200°C$ in Air

Items	N = 10000	N = 30000	N = 60000
ψ	0.2	0.2	0.2
Cracking mode 3	Case 2	Case 3	Case 4
Cracking mode 5	Case 2	Case 3	Case 4

fatigue tests were conducted at the loading frequency $f = 0.1$ and 1 Hz with a stress ratio $R = 0.05$.

Under $\sigma_{max} = 140$ MPa and $f = 0.1$ Hz, the experimental and theoretical hysteresis loops, interface slip of matrix cracking mode 3 and mode 5 corresponding to $N = 100, 1000,$ and 10000 are shown in Figures 3.24 through 3.26.

When $N = 100$, the fatigue hysteresis loops of matrix cracking mode 3 and mode 5, the composite and experimental data are shown in Figure 3.24a, in which the proportion of matrix cracking mode 3 is $\psi = 0.2$:

- For matrix cracking mode 3, the hysteresis loops correspond to interface slip Case 2, as shown in Figure 3.24b. Upon unloading to the valley stress, the interface counter-slip length approaches 54.6% of the interface debonding length, that is, $L_y(\sigma_{min})/L_d = 54.6\%$, and upon reloading to the peak stress, the interface new-slip length approaches 54.6% of the interface debonding length, that is, $L_z(\sigma_{max})/L_d = 54.6\%$.

- For matrix cracking mode 5, the hysteresis loops correspond to the interface slip Case 2, as shown in Figure 3.24b. Upon unloading to the valley stress, the interface counter-slip length approaches 72.8% of the interface debonding length, that is, $L_y(\sigma_{min})/L_d = 72.8\%$, and upon reloading to the peak stress, the interface new-slip length approaches 72.8% of the interface debonding length, that is, $L_z(\sigma_{max})/L_d = 72.8\%$.

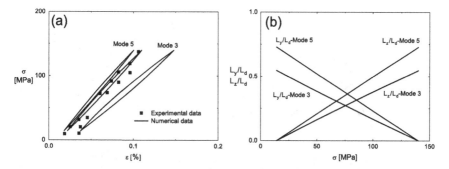

FIGURE 3.24 (a) Experimental and predicted hysteresis loops and (b) the interface slip lengths of matrix cracking modes 3 and 5 of 2D SiC/SiC composite under $\sigma_{max} = 140$ MPa at $N = 100$.

When N = 1000, the fatigue hysteresis loops of matrix cracking mode 3 and mode 5, the composite and experimental data are shown in Figure 3.25a, in which the proportion of matrix cracking mode 3 is ψ = 0.2:

- For matrix cracking mode 3, the hysteresis loops correspond to interface slip Case 4, as shown in Figure 3.25b. Upon unloading, the interface counter-slip length approaches to the interface debonding length at σ_{tr_fu} = 39.2 MPa, that is, $L_y(\sigma_{tr\_fu})/L_d$ = 1, and upon reloading to σ_{tr_fr} = 114.8 MPa, the interface new-slip length approaches to the interface debonding length, that is, $L_z(\sigma_{tr\_fr})/L_d$ = 1.

- For matrix cracking mode 5, the hysteresis loops correspond to the interface slip Case 2, as shown in Figure 3.25b. Upon unloading to the valley stress, the interface counter-slip length approaches 71.3% of the interface debonding length, that is, $L_y(\sigma_{min})/L_d$ = 71.3%, and upon reloading to the peak stress, the interface new-slip length approaches to 71.3% of the interface debonding length, that is, $L_z(\sigma_{max})/L_d$ = 71.3%.

When N = 10000, the fatigue hysteresis loops of matrix cracking mode 3 and mode 5, the composite and experimental data are shown in Figure 3.26(a), in which the proportion of matrix cracking mode 3 is ψ = 0.2:

- For matrix cracking mode 3, the hysteresis loops correspond to interface slip Case 4, as shown in Figure 3.26b. Upon unloading, the interface counter-slip length approaches the interface debonding

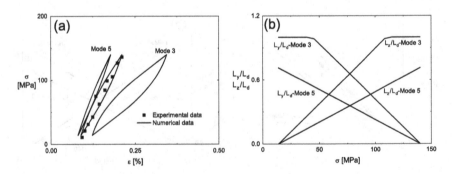

FIGURE 3.25 (a) Experimental and predicted hysteresis loops and (b) the interface slip lengths of matrix cracking modes 3 and 5 of 2D SiC/SiC composite under σ_{max} = 140 MPa at N = 1000.

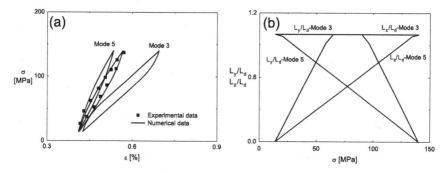

FIGURE 3.26 (a) Experimental and predicted hysteresis loops and (b) the interface slip lengths of matrix cracking modes 3 and 5 of 2D SiC/SiC composite under σ_{max} = 140 MPa at N = 10000.

length at σ_{tr_fu} = 89.6 MPa, that is, $L_y(\sigma_{tr\_fu})/L_d$ = 1, and upon reloading to σ_{tr_fr} = 64.4 MPa, the interface new-slip length approaches the interface debonding length, that is, $L_z(\sigma_{tr\_fr})/L_d$ = 1.

- For matrix cracking mode 5, the hysteresis loops correspond to the interface slip Case 4, as shown in Figure 3.26b. Upon unloading, the interface counter-slip length approaches to the interface debonding length at σ_{tr_fu} = 14 MPa, that is, $L_y(\sigma_{tr\_fu})/L_d$ = 1, and upon reloading to σ_{tr_fr} = 140 MPa, the interface new-slip length approaches to the interface debonding length, that is, $L_z(\sigma_{tr\_fr})/L_d$ = 1.

When σ_{max} = 140 MPa and f = 1 Hz, the experimental and theoretical hysteresis loops, the interface slip of matrix cracking modes 3 and 5 corresponding to N = 1000, 10000, and 30000 are shown in Figures 3.27 through 3.29.

When N = 1000, the fatigue hysteresis loops of matrix cracking modes 3 and 5, the composite and experimental data are shown in Figure 3.27a, in which the proportion of matrix cracking mode 3 is ψ = 0.2:

- For matrix cracking mode 3, the hysteresis loops correspond to interface slip Case 2, as shown in Figure 3.27b. Upon unloading to the valley stress, the interface counter-slip length approaches 53.9% of the interface debonding length, that is, $L_y(\sigma_{min})/L_d$ = 53.9%, and upon reloading to the peak stress, the interface new-slip length approaches 53.9% of the interface debonding length, that is, $L_z(\sigma_{max})/L_d$ = 53.9%.

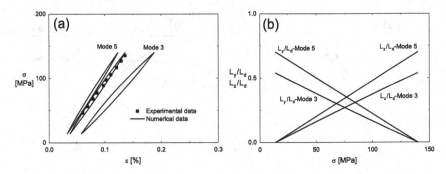

FIGURE 3.27 (a) Experimental and predicted hysteresis loops and (b) the interface slip lengths of matrix cracking modes 3 and 5 of 2D SiC/SiC composite under σ_{max} = 140 MPa at N = 1000.

- For matrix cracking mode 5, the hysteresis loops correspond to interface slip Case 2, as shown in Figure 3.27b. Upon unloading to the valley stress, the interface counter-slip length approaches to 70.2% of the interface debonding length, that is, $L_y(\sigma_{min})/L_d$ = 70.2%, and upon reloading to the peak stress, the interface new-slip length approaches 70.2% of the interface debonding length, that is, $L_z(\sigma_{max})/L_d$ = 70.2%.

When N = 10000, the fatigue hysteresis loops of matrix cracking mode 3 and mode 5, the composite and experimental data are given in Figure 3.28a, in which the proportion of matrix cracking mode 3 is ψ = 0.2:

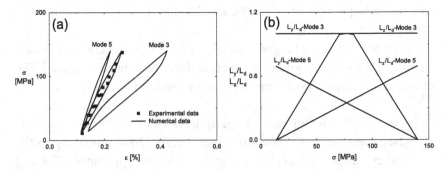

FIGURE 3.28 (a) Experimental and predicted hysteresis loops and (b) the interface slip lengths of matrix cracking modes 3 and 5 of 2D SiC/SiC composite under σ_{max} = 140 MPa at N = 10000.

- For matrix cracking mode 3, the hysteresis loops correspond to interface slip Case 4, as shown in Figure 3.28b. Upon unloading, the interface counter-slip length approaches the interface debonding length at $\sigma_{tr\_fu} = 77$ MPa, that is, $L_y(\sigma_{tr\_fu})/L_d = 1$, and upon reloading to $\sigma_{tr\_fr} = 77$ MPa, the interface new-slip length approaches the interface debonding length, that is, $L_z(\sigma_{tr\_fr})/L_d = 1$.

- For matrix cracking mode 5, the hysteresis loops correspond to the interface slip Case 2, as shown in Figure 3.28b. Upon unloading to the valley stress, the interface counter-slip length approaches 69.3% of the interface debonding length, that is, $L_y(\sigma_{min})/L_d = 69.3\%$, and upon reloading to the peak stress, the interface new-slip length approaches 69.3% of the interface debonding length, that is, $L_z(\sigma_{max})/L_d = 69.3\%$.

When $N = 30000$, the fatigue hysteresis loops of matrix cracking mode 3 and mode 5, the composite and experimental data are shown in Figure 3.29a, in which the proportion of matrix cracking mode 3 is $\psi = 0.2$:

- For matrix cracking mode 3, the hysteresis loops correspond to interface slip Case 4, as shown in Figure 3.29b. Upon unloading, the interface counter-slip length approaches the interface debonding length at $\sigma_{tr\_fu} = 108.5$ MPa, that is, $L_y(\sigma_{tr\_fu})/L_d = 1$, and upon reloading to $\sigma_{tr\_fr} = 45.5$ MPa, the interface new-slip length approaches the interface debonding length, that is, $L_z(\sigma_{tr\_fr})/L_d = 1$.

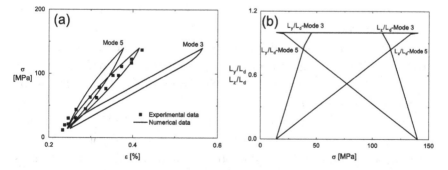

FIGURE 3.29 (a) Experimental and predicted hysteresis loops and (b) the interface slip lengths of matrix cracking modes 3 and 5 of 2D SiC/SiC composite under $\sigma_{max} = 140$ MPa at $N = 30000$.

- For matrix cracking mode 5, the hysteresis loops correspond to the interface slip Case 4, as shown in Figure 3.29b. Upon unloading, the interface counter-slip length approaches the interface debonding length at $\sigma_{tr\_fu} = 14$ MPa, that is, $L_y(\sigma_{tr\_fu})/L_d = 1$, and upon reloading to $\sigma_{tr\_fr} = 140$ MPa, the interface new-slip length approaches the interface debonding length, that is, $L_z(\sigma_{tr\_fr})/L_d = 1$.

Under $\sigma_{max} = 140$ MPa at $T = 1200°C$ in a steam condition, the proportion of matrix cracking mode 3 occupies 20% of all matrix cracking modes in a 2D SiC/SiC composite. When the loading frequency is $f = 0.1$ Hz, the interface shear stress decreases from $\tau_i = 12$ MPa at $N = 100$ to $\tau_i = 3$ MPa at $N = 10000$, and the hysteresis loops of matrix cracking modes 3 and 5 correspond to the interface slip Case 2 and Case 2 when $N = 100$, Case 4 and Case 2 when $N = 1000$, and Case 4 and Case 4 when $N = 10000$, respectively. When the loading frequency is $f = 1$ Hz, the interface shear stress decreases from $\tau_i = 10$ MPa at $N = 1000$ to $\tau_i = 3$ MPa at $N = 30000$; and the hysteresis loops of matrix cracking modes 3 and 5 correspond to the interface slip Case 2 and Case 2 when $N = 1000$, Case 2 and Case 4 when $N = 10000$, and Case 4 and Case 4 when $N = 30000$, respectively.

3.3.5 Cyclic-Fatigue Hysteresis Loops at 1300°C in Air

Zhu et al. [7] investigated the tension-tension cyclic-fatigue behavior of 2D SiC/SiC composite at $T = 1300$ °C in air condition. The composite used in the tension-tension fatigue tests was processed by using CVI of SiC (Nicalon™) into plane-woven [0°/90°] SiC-fiber preforms. Before infiltration, the preforms were coated with carbon by using chemical vapor deposition, to decrease the interface bonding between fibers and the matrix. The material contained $V_f = 40\%$ volume fraction of SiC fibers and had a porosity 9.7%. The monotonic tensile test was conducted under a constant displacement rate of 0.5 mm/min, and the monotonic tensile strength is 225 MPa. The tensile curve of 2D SiC/SiC composite at $T = 1300$ °C in air is shown in Figure 3.30.

Under $\sigma_{max} = 90$ MPa, the experimental and theoretical hysteresis loops, interface slip of matrix cracking modes 3 and 5 corresponding to $N = 6000, 24000, 90000, 650000, 1200000$, and 2800000 are shown in Figures 3.31 through 3.36.

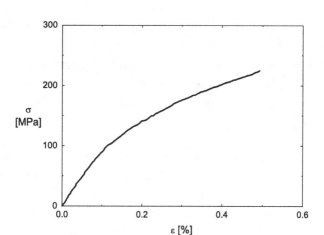

FIGURE 3.30   Monotonic tensile curve of 2D SiC/SiC composite at 1300°C in air.

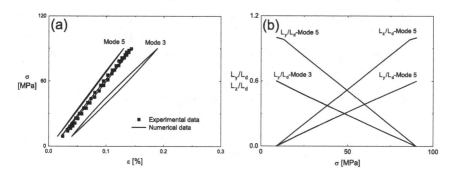

FIGURE 3.31   (a) Experimental and predicted hysteresis loops and (b) the inter-face slip lengths of matrix cracking modes 3 and 5 of 2D SiC/SiC composite under σ_{max} = 90 MPa at N = 6000.

When N = 6000, the fatigue hysteresis loops of matrix cracking mode 3 and mode 5, the composite and experimental data are shown in Figure 3.31a, in which the proportion of matrix cracking mode 3 is ψ = 0.25:

- For matrix cracking mode 3, the hysteresis loops correspond to interface slip Case 2, as shown in Figure 3.31b. Upon unloading to the valley stress, the interface counter-slip length approaches 60% of the interface debonding length, that is, $L_y(\sigma_{min})/L_d$ = 60%, and upon reloading to the peak stress, the interface new-slip length approaches 60% of the interface debonding length, that is, $L_z(\sigma_{max})/L_d$ = 60%.

- For matrix cracking mode 5, the hysteresis loops correspond to the interface slip Case 1. Upon unloading, the interface counter-slip length approaches the interface debonding length at $\sigma_{min} = 9$ MPa, i.e., $L_y(\sigma_{min})/L_d = 1$, and upon reloading to $\sigma_{max} = 90$ MPa, the interface new-slip length approaches to the interface debonding length, that is, $L_z(\sigma_{max})/L_d = 1$.

When $N = 24000$, the fatigue hysteresis loops of matrix cracking mode 3 and mode 5, the composite and experimental data are given in Figure 3.32a, in which the proportion of matrix cracking mode 3 is $\psi = 0.25$:

- For matrix cracking mode 3, the hysteresis loops correspond to interface slip Case 2, as shown in Figure 3.32b. Upon unloading to the valley stress, the interface counter-slip length approaches 59.9% of the interface debonding length, that is, $L_y(\sigma_{min})/L_d = 59.9\%$, and upon reloading to the peak stress, the interface new-slip length approaches 59.9% of the interface debonding length, that is, $L_z(\sigma_{max})/L_d = 59.9\%$.

- For matrix cracking mode 5, the hysteresis loops correspond to the interface slip Case 1. Upon unloading, the interface counter-slip length approaches the interface debonding length at $\sigma_{min} = 9$ MPa, that is, $L_y(\sigma_{min})/L_d = 1$, and upon reloading to $\sigma_{max} = 90$ MPa, the interface new-slip length approaches to interface debonded length, that is, $L_z(\sigma_{max})/L_d = 1$.

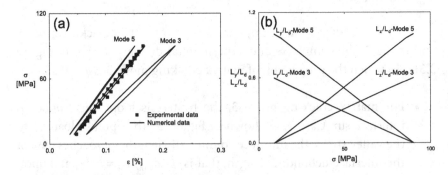

FIGURE 3.32 (a) Experimental and predicted hysteresis loops and (b) the interface slip lengths of matrix cracking modes 3 and 5 of 2D SiC/SiC composite under $\sigma_{max} = 90$ MPa at $N = 24000$.

When N = 90000, the fatigue hysteresis loops of matrix cracking mode 3 and mode 5, the composite and experimental data are given in Figure 3.33a, in which the proportion of matrix cracking mode 3 is ψ = 0.25:

- For matrix cracking mode 3, the hysteresis loops correspond to interface slip Case 2, as shown in Figure 3.33b. Upon unloading to the valley stress, the interface counter-slip length approaches 59.5% of the interface debonding length, that is, $L_y(\sigma_{min})/L_d$ = 59.5%, and upon reloading to the peak stress, the interface new-slip length approaches 59.5% of the interface debonding length, that is, $L_z(\sigma_{max})/L_d$ = 59.5%.

- For matrix cracking mode 5, the hysteresis loops correspond to the interface slip Case 2. Upon unloading to the valley stress, the interface counter-slip length approaches 99.5% of the interface debonding length, that is, $L_y(\sigma_{min})/L_d$ = 99.5%, and upon reloading to the peak stress, the interface new-slip length approaches 99.5% of the interface debonding length, that is, $L_z(\sigma_{max})/L_d$ = 99.5%.

When N = 650000, the fatigue hysteresis loops of matrix cracking mode 3 and mode 5, the composite and experimental data are shown in Figure 3.34a, in which the proportion of matrix cracking mode 3 is ψ = 0.25:

- For matrix cracking mode 3, the hysteresis loops correspond to interface slip Case 2, as shown in Figure 3.34b. Upon unloading to the valley stress, the interface counter-slip length approaches 59.1% of

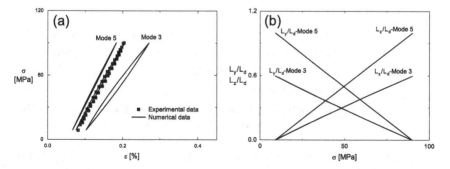

FIGURE 3.33 (a) Experimental and predicted hysteresis loops and (b) the interface slip lengths of matrix cracking modes 3 and 5 of 2D SiC/SiC composite under σ_{max} = 90 MPa at N = 90000.

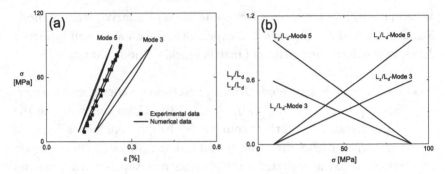

FIGURE 3.34   (a) Experimental and predicted hysteresis loops and (b) the interface slip lengths of matrix cracking modes 3 and 5 of 2D SiC/SiC composite under σ_{max} = 90 MPa at N = 650000.

the interface debonding length, that is, $L_y(\sigma_{min})/L_d$ = 59.1%; and upon reloading to the peak stress, the interface new-slip length approaches 59.1% of the interface debonding length, that is, $L_z(\sigma_{max})/L_d$ = 59.1%.

- For matrix cracking mode 5, the hysteresis loops correspond to the interface slip Case 2. Upon unloading to the valley stress, the interface counter-slip length approaches 97% of the interface debonding length, i.e., $L_y(\sigma_{min})/L_d$ = 97%, and upon reloading to the peak stress, the interface new-slip length approaches 97% of the interface debonding length, that is, $L_z(\sigma_{max})/L_d$ = 97%.

When N = 1200000, the fatigue hysteresis loops of matrix cracking mode 3 and mode 5, the composite and experimental data are shown in Figure 3.35a, in which the proportion of matrix cracking mode 3 is ψ = 0.25.

- For matrix cracking mode 3, the hysteresis loops correspond to interface slip Case 3, as shown in Figure 3.35b. Upon unloading to the valley stress, the interface counter-slip length approaches 70.4% of the interface debonding length, that is, $L_y(\sigma_{min})/L_d$ = 70.4%, and upon reloading to the peak stress, the interface new-slip length approaches 70.4% of the interface debonding length, that is, $L_z(\sigma_{max})/L_d$ = 70.4%.

- For matrix cracking mode 5, the hysteresis loops correspond to the interface slip Case 2. Upon unloading to the valley stress, the interface counter-slip length approaches 96.2% of the interface debonding

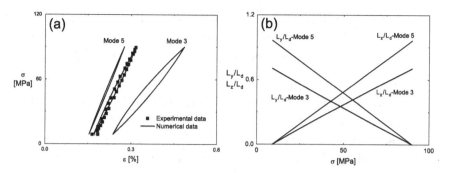

FIGURE 3.35　(a) Experimental and predicted hysteresis loops and (b) the interface slip lengths of matrix cracking modes 3 and 5 of 2D SiC/SiC composite under $\sigma_{max} = 90$ MPa at $N = 1200000$.

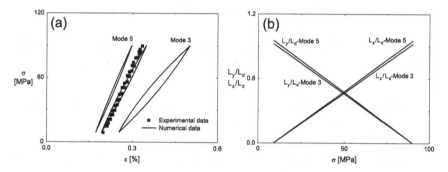

FIGURE 3.36　(a) Experimental and predicted hysteresis loops and (b) the interface slip lengths of matrix cracking modes 3 and 5 of 2D SiC/SiC composite under $\sigma_{max} = 90$ MPa at $N = 2800000$.

length, that is, $L_y(\sigma_{min})/L_d = 96.2\%$, and upon reloading to the peak stress, the interface new-slip length approaches 96.2% of the interface debonding length, that is, $L_z(\sigma_{max})/L_d = 96.2\%$.

When $N = 2800000$, the fatigue hysteresis loops of matrix cracking mode 3 and mode 5, the composite and experimental data are given in Figure 3.36a, in which the proportion of matrix cracking mode 3 is $\psi = 0.25$:

- For matrix cracking mode 3, the hysteresis loops correspond to interface slip Case 3, as shown in Figure 3.36b. Upon unloading to the valley stress, the interface counter-slip length approaches 92.3% of

the interface debonding length, that is, $L_y(\sigma_{min})/L_d = 92.3\%$, and upon reloading to the peak stress, the interface new-slip length approaches 92.3% of the interface debonding length, that is, $L_z(\sigma_{max})/L_d = 92.3\%$.

- For matrix cracking mode 5, the hysteresis loops correspond to the interface slip Case 2. Upon unloading to the valley stress, the interface counter-slip length approaches 95.4% of the interface debonding length, that is, $L_y(\sigma_{min})/L_d = 95.4\%$, and upon reloading to the peak stress, the interface new-slip length approaches 95.4% of the interface debonding length, that is, $L_z(\sigma_{max})/L_d = 95.4\%$.

When $\sigma_{max} = 120$ MPa, the experimental and theoretical hysteresis loops, interface slip of matrix cracking modes 3 and 5 corresponding to $N = 100$, 6000, 18000, and 36000 are shown in Figures 3.37 through 3.40.

When $N = 100$, the fatigue hysteresis loops of matrix cracking mode 3 and mode 5, the composite and experimental data are shown in Figure 3.37a, in which the proportion of matrix cracking mode 3 is $\psi = 0.4$:

- For matrix cracking mode 3, the hysteresis loops correspond to interface slip Case 2, as shown in Figure 3.37b. Upon unloading to the valley stress, the interface counter-slip length approaches 55.9% of the interface debonding length, that is, $L_y(\sigma_{min})/L_d = 55.9\%$; and upon reloading to the peak stress, the interface new-slip length approaches 55.9% of the interface debonding length, that is, $L_z(\sigma_{max})/L_d = 55.9\%$.

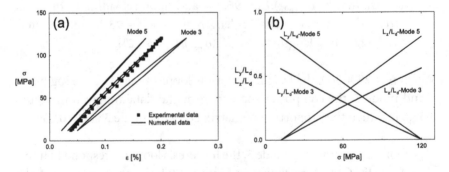

FIGURE 3.37　(a) Experimental and predicted hysteresis loops and (b) the interface slip lengths of matrix cracking modes 3 and 5 of 2D SiC/SiC composite under $\sigma_{max} = 120$ MPa at $N = 100$.

- For matrix cracking mode 5, the hysteresis loops correspond to the interface slip Case 2. Upon unloading to the valley stress, the interface counter-slip length approaches 80.7% of the interface debonding length, that is, $L_y(\sigma_{min})/L_d$ = 80.7%, and upon reloading to the peak stress, the interface new-slip length approaches 80.7% of the interface debonding length, that is, $L_z(\sigma_{max})/L_d$ = 80.7%.

When N = 6000, the fatigue hysteresis loops of matrix cracking mode 3 and mode 5, the composite and experimental data are given in Figure 3.38a, in which the proportion of matrix cracking mode 3 is ψ = 0.4:

- For matrix cracking mode 3, the hysteresis loops correspond to interface slip Case 2, as shown in Figure 3.38b. Upon unloading to the valley stress, the interface counter-slip length approaches 55.2% of the interface debonding length, that is, $L_y(\sigma_{min})/L_d$ = 55.2%; and upon reloading to the peak stress, the interface new-slip length approaches 55.2% of the interface debonding length, that is, $L_z(\sigma_{max})/L_d$ = 55.2%.

- For matrix cracking mode 5, the hysteresis loops correspond to the interface slip Case 2. Upon unloading to the valley stress, the interface counter-slip length approaches 77.2% of the interface debonding length, that is, $L_y(\sigma_{min})/L_d$ = 77.2%, and upon reloading to the peak stress, the interface new-slip length approaches 77.2% of the interface debonding length, that is, $L_z(\sigma_{max})/L_d$ = 77.2%.

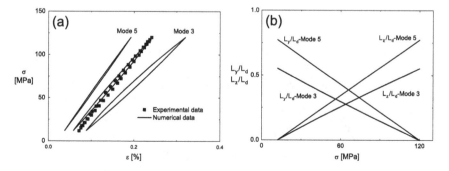

FIGURE 3.38 (a) Experimental and predicted hysteresis loops and (b) the interface slip lengths of matrix cracking modes 3 and 5 of 2D SiC/SiC composite under σ_{max} = 120 MPa at N = 6000.

When N = 18000, the fatigue hysteresis loops of matrix cracking mode 3 and mode 5, the composite and experimental data are given in Figure 3.39a, in which the proportion of matrix cracking mode 3 is ψ = 0.4:

- For matrix cracking mode 3, the hysteresis loops correspond to interface slip Case 2, as shown in Figure 3.39b. Upon unloading to the valley stress, the interface counter-slip length approaches 55% of the interface debonding length, that is, $L_y(\sigma_{min})/L_d$ = 55%; and upon reloading to the peak stress, the interface new-slip length approaches 55% of the interface debonding length, that is, $L_z(\sigma_{max})/L_d$ = 55%.

- For matrix cracking mode 5, the hysteresis loops correspond to the interface slip Case 2. Upon unloading to the valley stress, the interface counter-slip length approaches 76.4% of the interface debonding length, that is, $L_y(\sigma_{min})/L_d$ = 76.4%, and upon reloading to the peak stress, the interface new-slip length approaches 76.4% of the interface debonding length, that is, $L_z(\sigma_{max})/L_d$ = 76.4%.

When N = 36000, the fatigue hysteresis loops of matrix cracking mode 3 and mode 5, the composite and experimental data are given in Figure 3.40a, in which the proportion of matrix cracking mode 3 is ψ = 0.4:

- For matrix cracking mode 3, the hysteresis loops correspond to interface slip Case 3, as shown in Figure 3.40b. Upon unloading to the valley stress, the interface counter-slip length approaches 61.5% of the interface debonding length, that is, $L_y(\sigma_{min})/L_d$ = 61.5%, and upon

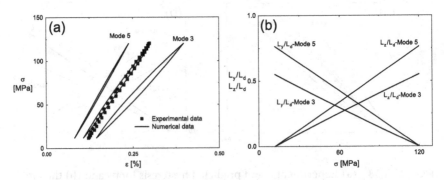

FIGURE 3.39 (a) Experimental and predicted hysteresis loops and (b) the interface slip lengths of matrix cracking modes 3 and 5 of 2D SiC/SiC composite under σ_{max} = 120 MPa at N = 18000.

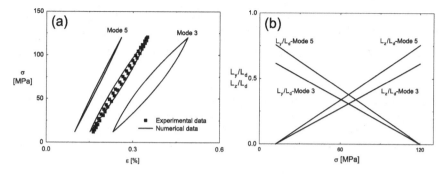

FIGURE 3.40 (a) Experimental and predicted hysteresis loops and (b) the interface slip lengths of matrix cracking modes 3 and 5 of 2D SiC/SiC composite under σ_{max} = 120 MPa at N = 36000.

reloading to the peak stress, the interface new-slip length approaches 61.5% of the interface debonding length, that is, $L_z(\sigma_{max})/L_d = 61.5\%$.

- For matrix cracking mode 5, the hysteresis loops correspond to the interface slip Case 2. Upon unloading to the valley stress, the interface counter-slip length approaches 75.6% of the interface debonding length, that is, $L_y(\sigma_{min})/L_d = 75.6\%$, and upon reloading to peak stress, the interface new-slip length approaches 75.6% of the interface debonding length, that is, $L_z(\sigma_{max})/L_d = 75.6\%$.

Under fatigue peak stress of $\sigma_{max} = 0.4\sigma_{uts}$ at T = 1300 °C in air, the proportion of matrix cracking mode 3 occupies 25% of all matrix cracking modes in 2D SiC/SiC composite; interface shear stress decreases from τ_i = 12 MPa at N = 6000 to τ_i = 3 MPa at N = 2800000; and the hysteresis loops of matrix cracking modes 3 and 5 correspond to the interface slip Case 2 and Case 1 when N = 6000, Case 2 and Case 1 when N = 24000, Case 2 and Case 2 when N = 90000, Case 2 and Case 2 when N = 650000, Case 3 and Case 2 when N = 1200000, and Case 3 and Case 2 when N = 2800000, as shown in Table 3.5. Under fatigue peak stress of $\sigma_{max} = 0.53\sigma_{uts}$ at 1300°C in air, the proportion of matrix cracking mode 3 occupies 40% of all matrix cracking modes in 2D SiC/SiC composite, and interface shear stress decreases from τ_i = 18 MPa at N = 100 to τ_i = 6 MPa at N = 36000, and the hysteresis loops of matrix cracking modes 3 and 5 correspond to the interface slip Case 2 and Case 2 when N = 100, Case 2 and Case 2 when N = 6000, Case 2 and Case 2 when N = 18000, and Case 3 and Case 2 when N = 36000, as shown in Table 3.6.

TABLE 3.5 The Damage Parameter and Interface Slip Type of 2D SiC/SiC Composite under σ_{max} = 90 MPa and f = 20 Hz at T = 1300°C in Air

Items	N = 6000	N = 24000	N = 90000	N = 650000	N = 1200000	N = 2800000
ψ	0.25	0.25	0.25	0.25	0.25	0.25
Cracking mode 3	Case 2	Case 2	Case 2	Case 2	Case 3	Case 3
Cracking mode 5	Case 1	Case 1	Case 2	Case 2	Case 2	Case 2

TABLE 3.6 The Damage Parameter and Interface Slip Type of 2D SiC/SiC Composite under σ_{max} = 120 MPa and f = 20 Hz at T = 1300 °C in Air

Items	N = 100	N = 6000	N = 18000	N = 36000
ψ	0.4	0.4	0.4	0.4
Cracking mode 3	Case 2	Case 2	Case 2	Case 3
Cracking mode 5	Case 2	Case 2	Case 2	Case 2

3.4 DISCUSSION

At elevated temperatures in air condition, matrix cracks would serve as avenues for the ingress of environment atmosphere into the composite. When the oxidizing gas ingresses into the composite, a sequence of events is triggered starting first with the oxidation of the interphase, then fibers. With increasing cycle number or oxidation time, the oxidation region propagates, and the interface shear stress decreases. The test temperature, loading frequency, and peak stress would affect the oxidation behavior, internal damage extent, and the fatigue hysteresis loops of 2D woven SiC/SiC composite.

When the test temperature increases from T = 1000 to 1200 °C in air, the damage evolution inside of 2D plain-woven SiC/SiC composite accelerated, for instance, (1) the degradation rate of interface shear stress increases, that is, the interface shear stress decreases from τ_i = 15 MPa at N = 2 to τ_i = 10 MPa at N = 30000 under σ_{max} = 0.58σ_{uts} at T = 1000°C, and the interface shear stress decreases from τ_i = 15 MPa at N = 1000 to τ_i = 2 MPa at N = 30000 under σ_{max} = 0.45σ_{uts} at T = 1200 °C; (2) the interface slip range increases, that is, the hysteresis loops of matrix cracking modes 3 and 5 both correspond to interface partially slip from N = 2 to 30000 under $\sigma_{max} = 0.58\sigma_{uts}$ at $T = 1000$ °C, the interface partially slips at $N = 1000$, and the interface completely slips at $N = 30000$ under $\sigma_{max} = 0.45\sigma_{uts}$ at $T = 1200$ °C.

When the loading frequency decreases from $f = 1.0$ to 0.1 Hz at $T = 1200$°C in air, the damage evolution inside 2D plain-woven SiC/SiC composite accelerated, for instance, (1) the degradation rate of interface shear stress increases, that is, from $\tau_i = 15$ MPa at $N = 1000$ to $\tau_i = 2$ MPa at $N = 30000$ when $f = 0.1$ Hz, and from $\tau_i = 10$ MPa at $N = 10000$ to $\tau_i = 2$ MPa at $N = 60000$ when $f = 1.0$ Hz, and (2) the interface slip range increases, that is, the hysteresis loops of matrix cracking modes 3 and 5 correspond to the interface partially slip at $N = 1000$, the interface completely slips at $N = 30000$ when $f = 0.1$ Hz, and interface partially slip at $N = 10000$, and completely slip at $N = 60000$ when $f = 1.0$ Hz.

When the fatigue peak stress increases from $\sigma_{max} = 0.4\sigma_{uts}$ to $0.53\sigma_{uts}$ at $T = 1300$ °C in air, the damage evolution inside 2D plain-woven SiC/SiC composite accelerated, for instance, (1) the proportion of matrix cracking mode 3 increases from $\psi = 0.25$ to 0.4; (2) the degradation rate of interface shear stress increases, that is, from $\tau_i = 12$ MPa at $N = 6000$ to $\tau_i = 3$ MPa at $N = 2800000$ when $\sigma_{max} = 0.4\sigma_{uts}$ and from $\tau_i = 18$ MPa at $N = 100$ to $\tau_i = 6$ MPa at $N = 36000$ when $\sigma_{max} = 0.53\sigma_{uts}$; and (3) the interface slip range increases, that is, the hysteresis loops of matrix cracking modes 3 and 5 correspond to interface partially slip at $N = 6000$, interface completely and partially slips at $N = 2800000$ when $\sigma_{max} = 0.4\sigma_{uts}$, the interface partially slips at $N = 100$, and interface completely and partially slip at $N = 36000$ when $\sigma_{max} = 0.53\sigma_{uts}$.

3.5 SUMMARY AND CONCLUSION

In this chapter, the cyclic fatigue hysteresis loops of 2D plain-woven SiC/SiC composite at $T = 1000$, 1200, and 1300°C in air and in steam conditions have been investigated. The interface slip between fibers and the matrix existed in the matrix cracking modes 3 and 5 were considered to be the major reason for the hysteresis loops of 2D plain-woven CMCs. The hysteresis loops and interface slip of 2D SiC/SiC composite corresponding to different fatigue peak stresses, test conditions, and loading frequencies have been predicted using the present analysis. The damage evolution inside of 2D woven SiC/SiC composite accelerated with an increase in test temperature and fatigue peak stress and a decrease of loading frequency at elevated temperatures in an air condition, due to interphase and fibers oxidation at elevated temperatures in an air condition.

(1) At $T = 1000$, 1200, and 1300 °C in air, the interface shear stress in the longitudinal yarns decreases with increasing cycle number due to interphase oxidation and interface wear, that is, from $\tau_i = 15$ MPa at $N = 1000$ to $\tau_i = 2$ MPa at $N = 30000$ under $\sigma_{max} = 0.45\sigma_{uts}$ at $T = 1200$ °C, leading to the transition of interface slip type of matrix cracking mode 3 and mode 5, that is, from the interface partially slips at $N = 1000$ to interface completely slip at $N = 30000$ under $\sigma_{max} = 0.45\sigma_{uts}$ at $T = 1200$ °C.

(2) The damage parameter, that is, the proportion of matrix cracking mode 3 in the entire matrix cracking modes of the composite and the fatigue hysteresis dissipated energy increased with increasing fatigue peak stress.

REFERENCES

1. Gowayed Y, Ojard G, Santhosh U, Jefferso G. Modeling of crack density in ceramic matrix composites. *J. Compos. Mater.* 2015; 49:2285–2294.
2. Evans AG, Zok FW, McMeeking RM. Fatigue of ceramic matrix composites. *Acta Metal. Mater.* 1995; 43:859–875.
3. Reynaud P. Cyclic fatigue of ceramic-matrix composites at ambient and elevated temperatures. *Compos. Sci. Technol.* 1996; 56:809–814.
4. Fantozzi G, Reynaud P. Mechanical hysteresis in ceramic matrix composites. *Mater. Sci. Eng. Part A.* 2009; 521–522:18–23.
5. Michael K. Fatigue behavior of a SiC/SiC composite at 1000°C in air and steam. AFIT/GAE/ENY/10-D01, 2010.
6. Jacob D. Fatigue behavior of an advanced SiC/SiC composite with an oxidation inhibited matrix at 1200°C in air and in steam. AFIT/GEA/ENY/10-M07, 2010.
7. Zhu SJ, Mizuno M, Nagano Y, Cao JW, Kagawa Y, Kaya H. Creep and fatigue behavior in an enhanced SiC/SiC composite at high temperature. *J. Am. Ceram. Soc.* 1998; 81:2269–2277.
8. Kuo WS, Chou TW. Multiple cracking of unidirectional and cross-ply ceramic matrix composites. *J. Am. Ceram. Soc.* 1995; 78:745–755.
9. Li LB. Modeling cyclic fatigue hysteresis loops of 2D woven ceramic-matrix composite at elevated temperatures in air considering multiple matrix cracking modes. *Theor. Appl. Fract. Mech.* 2016; 85:246–261.

CHAPTER 4

High-Temperature Cyclic-Fatigue Mechanical Hysteresis Behavior in 2.5-Dimensional Woven SiC/SiC Composites

4.1 INTRODUCTION

Oxidation is the key factor in limiting the application of CMCs on hot section load-carrying components of aeroengines. Combining carbides deposited by the chemical vapor infiltration (CVI) process with specific sequences, a new generation of SiC/SiC composite with a self-healing matrix has been developed to improve the oxidation resistance [1, 2]. The self-sealing matrix forms a glass with oxygen at high temperatures and consequently prevents oxygen diffusion inside the material. At low temperatures 650–1000°C in dry and wet oxygen atmospheres, the self-healing 2.5-dimensional (2.5D) Nicalon™ NL202 SiC/[Si-B-C] with a pyrocarbon (PyC) interphase exhibits a better oxidation resistance compared to SiC/SiC with PyC, due to the presence of boron compounds [3]. The fatigue lifetime duration in an air atmosphere at intermediate and high temperatures is considerably reduced beyond the elastic yield point. For the Nicalon™ SiC/[Si-B-C] composite, the elastic yield point is about $\sigma = 80$ MPa. The lifetime duration is about $t = 10$–20 h at $T = 873$ K and less than $t = 1$ h at $T = 1123$ K under $\sigma_{max} = 120$ MPa. For the self-healing Hi-Nicalon™ SiC/SiC composite, a duration $t = 1000$ h without failure is reached at $\sigma_{max} = 170$ MPa, and a duration longer than $t = 100$ h at $\sigma_{max} = 200$ MPa and $T = 873$ K is reached [4]. For the self-healing Hi-Nicalon™ SiC/[SiC-B$_4$C] composite, at $T = 1200$ °C, there is little influence on the fatigue performance at $f = 1.0$ Hz, but there is a noticeably degraded fatigue lifetime at $f = 0.1$ Hz with the presence of steam [5, 6]. An increase in temperature from $T = 1200$ to 1300°C slightly degrades the fatigue performance in an air atmosphere but not in a steam atmosphere [7]. The crack growth in the SiC fiber controls the fatigue lifetime of self-healing Hi-Nicalon™ SiC/[Si-B-C] at $T = 873$ K, and the fiber creep controls the fatigue lifetime of self-healing SiC/[Si-B-C] at $T = 1200$ °C [8]. The typical cyclic fatigue behavior of a self-healing Hi-Nicalon™ SiC/[Si-B-C] composite involves an initial decrease of the effective modulus to a minimum value, followed by a stiffening, and the time-to-the minimum modulus is in inverse proportion to the loading frequency [9]. The initial cracks within the longitudinal tows caused by interphase oxidation contribute to the initial decrease of modulus. The glass produced by the oxidation of the self-healing matrix may contribute to the stiffening of the composite either by sealing the cracks or by bonding the fiber to the matrix [10]. The damage evolution of self-healing Hi-Nicalon™ SiC/[Si-B-C] composite at elevated temperatures can be monitored using acoustic emission (AE) [11, 12]. The relationship between interface oxidation and AE energy under static fatigue loading at elevated temperatures has been developed [13]. However, at a high temperature, above $T = 1000$ °C, AE cannot be applied for cyclic fatigue damage monitoring. The complex fatigue damage mechanisms of self-healing CMCs affect damage evolution and lifetime. Hysteresis loops are related to cycle-dependent fatigue damage mechanisms [14–16]. The damage parameters derived from hysteresis loops have already been applied for analyzing the fatigue damage and fracturing of different non-oxide CMCs at elevated temperatures [17–21].

The objective of this chapter is to analyze cycle-dependent damage development in self-healing 2.5D woven Hi-Nicalon™ SiC/[Si-B-C] at $T = 600$°C and 1200°C using damage evolution models and parameters. The cycle-dependent damage parameters of internal friction, dissipated energy, Kachanov's damage parameter, and broken fiber fraction were obtained to analyze damage development in self-healing CMCs.

Relationships between cycle-dependent damage parameters and multiple fatigue damage mechanisms are established. Experimental fatigue damage evolution of each of the composites—self-healing Hi-Nicalon™ SiC/[Si-B-C] is predicted. Effects of fatigue peak stress and testing environment on the evolution of internal damage and final fracture are analyzed.

4.2 MATERIALS AND EXPERIMENTAL PROCEDURES

The Hi-Nicalon™ fibers (Nippon Carbon Co., Ltd., Tokyo, Japan) with an interphase of pyrolytic carbon (PyC) reinforced multilayered matrix [Si-B-C] were provided by SNECMA Propulsion Solide, Le Haillan, France. The self-healing Hi-Nicalon™ SiC/[Si-B-C] composite was fabricated using chemical vapor infiltration (CVI). The experimental results were performed and obtained by Penas [22]. The detailed information of materials and experimental procedures of 2.5D Hi-Nicalon™ SiC/[Si-B-C] composite at 600and 1200°C in an air atmosphere are shown in Table 4.1.

TABLE 4.1 Materials and Experimental Procedures of 2.5D Woven Hi-Nicalon™ SiC/[Si-B-C] Composite at 600 and 1200°C in an Air Atmosphere

Materials	
Fiber	Hi-Nicalon™
Interphase	Pyrolytic Carbon (PyC)
Matrix	Multilayer [Si-B-C]
Fiber preform	2.5D
Fiber volume/(%)	35
Fabrication method	Chemical Vapor Infiltration (CVI)
Manufacturer	SNECMA Propulsion Solide
Experimental Procedures	
Specimen shape	Dog bone–shaped
Specimen dimension	200 mm length
	5 mm thickness
	16 mm width
Testing machine	INSTRON Model 8502 servo hydraulic load-frame
Loading frequency/(Hz)	0.25
Maximum cycle number	1000000
Temperature/(°C)	600, 1200
Environment	air
$\sigma_{min}/\sigma_{max}$ (MPa)	0/200, −50/300 at 600°C
	0/170, −50/200 at 1200°C

The 2.5D fiber preform was consolidated by an interphase PyC deposited by CVI with a thickness of 0.1 μm to optimize the interphase properties and promote the desired pseudo-ductile behavior in the composite. The fiber volume was about 35%, and the porosity was about 10%.

4.3 MICROMECHANICAL HYSTERESIS CONSTITUTIVE MODEL

When the peak stress is higher than the first matrix cracking stress, under cyclic fatigue loading, multiple fatigue damage mechanisms of matrix cracking, interface debonding, wear and oxidation, and fiber fracturing occur [23–27]. Hysteresis loops appear and evolve with cycle number upon unloading and reloading due to internal multiple damages in CMCs. A unit cell is extracted from the damaged CMCs, as shown in Figure 4.1. The total length of the unit cell is half of a matrix crack spacing ($L_c/2$), and the interface debonding length between the space of matrix cracking is L_d. Upon unloading, the debonding zone is divvied into a counter-slip zone with the length of L_y and a slip zone of length $L_d - L_y$, as shown in Figure 4.1(a); and upon reloading, the debonding zone is divvied into a new slip zone with a length of L_z, a counter-slip region with a length of $L_y - L_z$, and a slip region with a length of $L_d - L_y$, as shown in Figure 4.1b.

Based on the interface debonding and slip state between the matrix crack spacing, the types of hysteresis loops can be divided into four cases, as shown in Table 4.2.

For the Case 1 and 2 in Table 4.2, the unloading and reloading composite hysteresis strain is a function of cycle-dependent unloading intact fiber stress ($\Phi_U(N)$), reloading intact fiber stress ($\Phi_R(N)$), interface shear stress ($\tau_i(N)$), interface debonding, and slip length ($L_d(N)$, $L_y(N)$, and $L_z(N)$). The

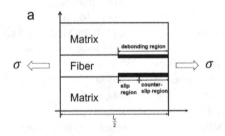

FIGURE 4.1   Unit cell of damaged CMCs on (a) unloading and (b) reloading.

TABLE 4.2   Interface Debonding and Slip State in CMCs

Case	Interface Debonding Condition	Interface Counter Slip Condition	Interface New Slip Condition
Case 1	$L_d(\sigma_{max}) < L_c/2$	$L_y(\sigma_{min}) = L_d(\sigma_{max})$	$L_z(\sigma_{max}) = L_d(\sigma_{max})$
Case 2	$L_d(\sigma_{max}) < L_c/2$	$L_y(\sigma_{min}) < L_d(\sigma_{max})$	$L_z(\sigma_{max}) < L_d(\sigma_{max})$
Case 3	$L_d(\sigma_{max}) = L_c/2$	$L_y(\sigma_{min}) < L_c/2$	$L_z(\sigma_{max}) < L_c/2$
Case 4	$L_d(\sigma_{max}) = L_c/2$	$L_y(\sigma_{min}) = L_c/2$	$L_z(\sigma_{max}) = L_c/2$

cycle-dependent unloading composite hysteresis strain ($\varepsilon_U(N)$) and reloading composite hysteresis strain ($\varepsilon_R(N)$) are determined by

$$\varepsilon_U(N) = \frac{\Phi_U(N)+\varphi(N)}{E_f} + 4\frac{\tau_i(N)}{E_f}\frac{L_y^2(N)}{r_f l_c(N)}$$
$$-\frac{\tau_i(N)}{E_f}\frac{\left(2L_y(N)-L_d(N)\right)\left(2L_y(N)+L_d(N)-L_c(N)\right)}{r_f L_c(N)} \quad (4.1)$$
$$-\left(\alpha_c - \alpha_f\right)\Delta T,$$

$$\varepsilon_R(N) = \frac{\Phi_R(N)+\varphi(N)}{E_f} - 4\frac{\tau_i(N)}{E_f}\frac{L_z^2(N)}{r_f L_c(N)}$$
$$+4\frac{\tau_i(N)}{E_f}\frac{\left(L_y(N)-2L_z(N)\right)^2}{r_f L_c(N)}$$
$$+2\frac{\tau_i(N)}{E_f}\frac{\left(L_d(N)+2L_y(N)-2L_z(N)-L_c(N)\right)\left(L_d(N)-2L_y(N)+2L_z(N)\right)}{r_f L_c(N)} \quad (4.2)$$
$$-\left(\alpha_c - \alpha_f\right)\Delta T,$$

where r_f is the fiber radius; α_f and α_c denote the fiber and composite's thermal expansion coefficient, respectively; and ΔT denotes the temperature difference between tested and fabricated temperatures.

For Cases 3 and 4 in Table 4.2, the cycle-dependent unloading composite hysteresis strain ($\varepsilon_U(N)$) and reloading composite hysteresis strain ($\varepsilon_R(N)$) can be expressed by

$$\varepsilon_U(N) = \frac{\Phi_U(N) + \Delta\varphi(N)}{E_f} + 4\frac{\tau_i(N)}{E_f}\frac{L_y^2(N)}{r_f L_c(N)}$$

$$-2\frac{\tau_i(N)}{E_f}\frac{\left(2L_y(N) - L_c(N)/2\right)^2}{r_f L_c(N)} - (\alpha_c - \alpha_f)\Delta T, \tag{4.3}$$

$$\varepsilon_R(N) = \frac{\Phi_R(N) + \Delta\varphi(N)}{E_f} - 4\frac{\tau_i(N)}{E_f}\frac{L_z^2(N)}{r_f l_c}$$

$$+4\frac{\tau_i(N)}{E_f}\frac{\left(L_y(N) - 2L_z(N)\right)^2}{r_f l_c}$$

$$-2\frac{\tau_i(N)}{E_f}\frac{\left(L_c(N)/2 - 2L_y(N) + 2L_z(N)\right)^2}{r_f L_c(N)} \tag{4.4}$$

$$-(\alpha_c - \alpha_f)\Delta T.$$

The cycle-dependent internal damage parameter is defined by $\Delta W/W_e$, where W_e is the maximum elastic energy stored during a cycle.

$$\Delta W(N) = \int_{\sigma_{min}}^{\sigma_{max}} \left[\varepsilon_U(N) - \varepsilon_R(N)\right]d\sigma \tag{4.5}$$

Substituting Equations 4.1 through 4.4 into Equation 4.5, the damage parameter ($\Delta W(N)$) can be obtained, which is a function of cycle-dependent unloading intact fiber stress ($\Phi_U(N)$); reloading intact fiber stress ($\Phi_R(N)$); interface shear stress ($\tau_i(N)$); interface debonding; and slip length, ($L_d(N)$, $L_y(N)$, and $L_z(N)$). It should be noted that the cycle-dependent unloading intact fiber stress ($\Phi_U(N)$) and reloading intact fiber stress ($\Phi_R(N)$) consider fiber failure and broken fiber fraction. Comparing experimental $\Delta W/W_e$ or ΔW with theoretical values, the interface shear stress ($\tau_i(N)$) and broken fiber fraction (P) can be obtained for different cycle numbers.

The mean elastic modulus (E) is the mean slope of the hysteresis loop. This modulus is usually normalized by Young's modulus E_0 of an undamaged composite, leading to the plotting of (E/E_0). A Kachanov's damage parameter (D) is given by

$$D = 1 - \frac{E}{E_0}. \qquad (4.6)$$

The Kachanov's damage parameter (D) is another way to describe the evolution of a composite's mean elastic modulus (E) under cyclic fatigue but contains the same information as the normalized modulus (E/E_0).

4.4 EXPERIMENTAL COMPARISONS

Monotonic tensile and cycle-dependent damage evolutions of self-healing 2.5D woven Hi-Nicalon™ SiC/[Si-B-C] were analyzed for different temperatures. The monotonic tensile curves exhibit obvious nonlinearity at elevated temperatures. For 2.5D woven Hi-Nicalon™ SiC/[Si-B-C] at T = 600 and 1200 °C, the tensile curves can be divided into three main zones. The cycle-dependent damage parameters of internal friction ($\Delta W(N)/W_e(N)$), dissipated energy ($\Delta W(N)$), interface shear stress ($\tau_i(N)$), Kachanov's damage parameter ($D(N)$), and broken fiber fraction ($P(N)$) versus cycle number are analyzed for different temperatures, peak stresses, and loading frequencies. The interface shear stress decreases with applied cycle number, and the Kachanov's damage parameter and broken fiber fraction increase with applied cycle number. However, the evolution of internal frictional and dissipated energy with more cycles is much more complex, as together, they depend on the peak stress, temperature, and testing environment. The internal damage evolution of 2.5D woven Hi-Nicalon™ SiC/[Si-B-C], when subjected to cyclic fatigue loading, was obtained.

4.4.1 2.5D Woven Hi-Nicalon™ SiC/[Si-B-C] at 600°C in an Air Atmosphere

The monotonic tensile curve of 2.5D woven self-healing Hi-Nicalon™ SiC/[Si-B-C] composite at T = 600°C in an air atmosphere is shown in Figure 4.2. The self-healing Hi-Nicalon™ SiC/[Si-B-C] composite fractures at σ_{UTS} = 341 MPa with the failure strain ε_f = 0.64%. The tensile curve exhibits obvious nonlinearity, and can be divided into three zones, including (1) the linear elastic zone with an elastic modulus E_c = 195 ± 20 GPa, (2) the nonlinear zone due to multiple matrix cracking, and (3) the second linear zone after saturation of matrix cracking up to final fracture, with an elastic modulus $E_c = 23 \pm 1$ GPa, which is half of the theoretical value $E_f V_{fl}$ (43 GPa) when the load is supported only by the longitudinal fiber.

Experimental cycle-dependent internal friction parameter ($\Delta W/W_e$) versus cycle number curves of 2.5D woven self-healing Hi-Nicalon™ SiC/[Si-B-C] composite under $\sigma_{min} = -50/\sigma_{max} = 300$ MPa and $\sigma_{min} = 0/\sigma_{max} = 200$ MPa at $T = 600°C$ in an air atmosphere are shown in Figure 4.3. The cycle-dependent internal friction parameter ($\Delta W/W_e$) decreases first,

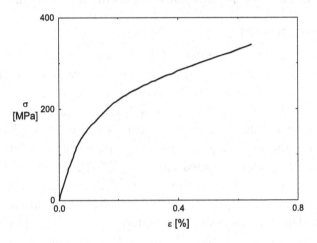

FIGURE 4.2 Monotonic tensile curve of 2.5D woven self-healing Hi-Nicalon™ SiC/[Si-B-C] composite at $T = 600°C$ in an air atmosphere.

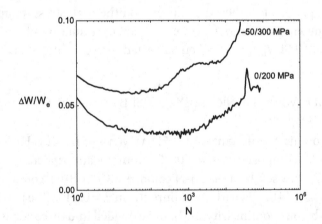

FIGURE 4.3 Experimental $\Delta W/W_e$ versus cycle number curves of 2.5D woven self-healing Hi-Nicalon™ SiC/[Si-B-C] composite at $T = 600°C$ in an air atmosphere.

followed by a short stabilization and increases again before reaching a plateau, and finally, there is a sharp increase when the composite approaches failure. During the initial stage of cyclic fatigue loading, matrix cracking and interface debonding occur when the fatigue peak stress is higher than the first matrix cracking stress. Under repeated unloading and reloading, the sliding between the fiber and the matrix leads to the interface wear and oxidation, which decreases the interface shear stress. The initial decrease of the internal friction parameter ($\Delta W/W_e$) is mainly attributed to matrix cracking, cycle-dependent interface debonding, and interface wear. However, with increasing cycle number, the interface wear and oxidation decrease the interface shear stress to a constant value, leading to the stabilization of cycle-dependent interface debonding and slip length and the internal friction damage parameter ($\Delta W/W_e$). The interface wear and oxidation also decrease the fiber strength, leading to the gradual fracture of fiber, and the sudden increase of internal friction at the end of the test corresponding to fiber broken.

Experimental and predicted cycle-dependent internal friction parameter ($\Delta W(N)/W_e(N)$) and broken fiber fraction ($P(N)$) versus the interface shear stress curves, and the cycle-dependent Kachanov's damage parameter ($D(N)$) and the interface shear stress ($\tau_i(N)$) versus cycle number curves of 2.5D woven self-healing Hi-Nicalon™ SiC/[Si-B-C] at 600°C in an air atmosphere are shown in Figure 4.4 and Table 4.3.

Under σ_{max} = 200 and 300 MPa, the internal damage parameter ($\Delta W(N)/W_e(N)$) first decreases with the interface shear stress, mainly due to the interface wear and oxidation, and then increases with the interface shear stress, mainly due to the fiber broken, corresponding to the interface slip Case 4 in Table 4.2, as shown in Figure 4.4a.

Under σ_{max} = 300 MPa, the broken fiber fraction ($P(N)$) at higher interface shear stress is much higher than that under σ_{max} = 200 MPa, mainly due to higher peak stress, as shown in Figure 4.4b; and the Kachanov's damage parameter ($D(N)$) is also higher than that under σ_{max} = 200 MPa, as shown in Figure 4.4(c). The interface shear stress under σ_{max} = 300 MPa is also higher than that under σ_{max} = 200 MPa, mainly due to the scatter of interface shear stress or compressive stress of σ_{min} = −50 MPa acting on the composite.

Under σ_{max} = 200 MPa, the cycle-dependent damage parameter ($\Delta W/W_e$) decreases first, that is, from $\Delta W/W_e$ = 0.054 at τ_i = 15.7 MPa to $\Delta W/W_e$ = 0.034 at τ_i = 8.0 MPa, and then increases from $\Delta W/W_e$ = 0.034 at τ_i = 8.0 MPa to $\Delta W/W_e$ = 0.062 at τ_i = 6.0 MPa. The cycle-dependent

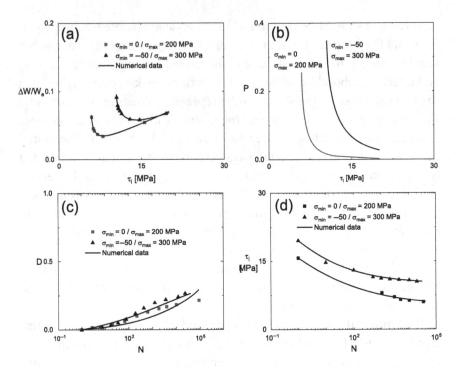

FIGURE 4.4 (a) Experimental and predicted $\Delta W/W_e$, (b) the broken fiber fraction (P), (c) the experimental and predicted Kachanov's damage parameter (D), and (d) the interface shear stress ($\tau_i(N)$) of 2.5D woven self-healing Hi-Nicalon™ SiC/[Si-B-C] composite at $T = 600°C$ in an air atmosphere.

broken fiber fraction (P) increases from $P = 0.004$ at $\tau_i = 15.7$ MPa to $P = 0.24$ at $\tau_i = 6.0$ MPa. The cycle-dependent Kachanov's damage parameter (D) increases from $D = 0$ at $N = 1$ to $D = 0.21$ at $N = 89459$. The cycle-dependent interface shear stress (τ_i) decreases from $\tau_i = 15.7$ MPa at $N = 1$ to $\tau_i = 6.0$ MPa at $N = 33788$.

Under the peak stress of $\sigma_{max} = 300$ MPa, the cycle-dependent damage parameter ($\Delta W/W_e$) decreases first, that is, from $\Delta W/W_e = 0.067$ at $\tau_i = 19.5$ MPa to $\Delta W/W_e = 0.058$ at $\tau_i = 14.7$ MPa, and then increases from $\Delta W/W_e = 0.058$ at $\tau_i = 14.7$ MPa to $\Delta W/W_e = 0.091$ at $\tau_i = 10.5$ MPa. The cycle-dependent broken fiber fraction (P) increases from $P = 0.029$ at $\tau_i = 19.5$ MPa to $P = 0.347$ at $\tau_i = 10.5$ MPa. The cycle-dependent Kachanov's damage parameter (D) increases from $D = 0$ at $N = 1$ to $D = 0.265$ at $N = 23666$. The interface shear stress (τ_i) decreases from $\tau_i = 19.5$ MPa at $N = 1$ to $\tau_i = 10.5$ MPa at $N = 19812$.

TABLE 4.3   Cycle-Dependent Damage Evolution of 2.5D Woven Self-healing Hi-Nicalon™ SiC/[Si-B-C] Composite at $T = 600°C$ in an air atmosphere

Cycle number	$\Delta W/W_e$	τ_i/MPa	P/%
0/200 MPa at 600°C in Air Atmosphere			
1	0.054	15.7	0.4
1055	0.034	8.0	4.5
2993	0.037	7.1	7.7
5041	0.042	6.5	12.3
10,343	0.045	6.3	15.1
33,788	0.062	6	24.0
−50/300 MPa at 600°C in Air Atmosphere			
1	0.067	19.5	2.9
10	0.058	14.7	7.0
100	0.059	13.0	10.9
500	0.067	11.5	18.4
1000	0.071	11.2	21.1
2080	0.074	11.0	23.5
4410	0.075	10.9	24.9
10,406	0.079	10.8	26.7
19,812	0.091	10.5	34.7

4.4.2 2.5D Woven Hi-Nicalon™ SiC/[Si-B-C] at 1200°C in an Air Atmosphere

The monotonic tensile curve of 2.5D woven self-healing Hi-Nicalon™ SiC/[Si-B-C] composite at $T = 1200°C$ in an air atmosphere is shown in Figure 4.5. The composite tensile fractured at $\sigma_{UTS} = 354$ MPa with $\varepsilon_f = 0.699\%$. The tensile curve can also be divided into three zones, including (1) the initial linear elastic zone, (2) the nonlinear zone, and (3) the second linear region with fiber broken. The average fracture strength and failure strain of 2.5D woven Hi-Nicalon™ SiC/[Si-B-C] composite are slightly lower at $T = 1200°C$; that is, $\sigma_{UTS} = 320$ MPa and $\varepsilon_f = 0.62\%$ against $\sigma_{UTS} = 332$ MPa and $\varepsilon_f = 0.658\%$ at $T = 600°C$.

The experimental cycle-dependent internal friction parameter ($\Delta W/W_e$) versus cycle number curves of 2.5D woven self-healing Hi-Nicalon™ SiC/[Si-B-C] at $T = 1200°C$ in an air atmosphere are shown in Figure 4.6. The cycle-dependent internal friction parameter ($\Delta W/W_e$) decreases continuously, and finally, there is a sharp increase when the composite approaches failure. The internal friction decreases as the interface wear reduces the

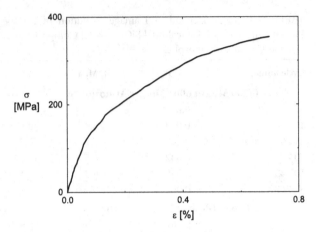

FIGURE 4.5 Tensile curve of 2.5D woven self-healing Hi-Nicalon™ SiC/[Si-B-C] composite at $T = 1200°C$ in an air atmosphere.

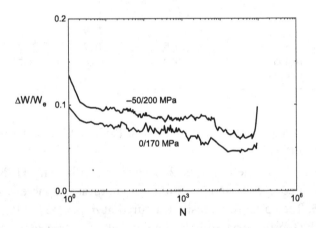

FIGURE 4.6 Experimental internal friction parameter ($\Delta W/W_e$) versus cycle number curves of 2.5D woven self-healing Hi-Nicalon™ SiC/[Si-B-C] composite at $T = 1200°C$ in an air atmosphere.

interface shear stress. The sudden increase of internal friction at the end of the test corresponds to the fiber broken.

Experimental and predicted cycle-dependent internal friction parameter ($\Delta W(N)/W_e(N)$) and the broken fiber fraction ($P(N)$) versus the interface shear stress curves; and the cycle-dependent Kachanov's damage parameter ($D(N)$) and the interface shear stress ($\tau_i(N)$) versus cycle number curves of 2.5D woven self-healing Hi-Nicalon™ SiC/[Si-B-C] composite at 1200°C in an air atmosphere are shown in Figure 4.7 and Table 4.4.

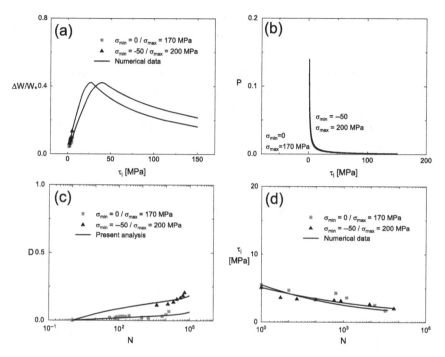

FIGURE 4.7 (a) Experimental and predicted internal friction parameter $(\Delta W/W_e)$, (b) the broken fiber fraction (P), (c) the experimental and predicted Kachanov's damage parameter (D), and (d) the interface shear stress of 2.5D woven self-healing Hi-Nicalon™ SiC/[Si-B-C] composite at $T = 1200°C$ in an air atmosphere.

Under $\sigma_{max} = 170$ and 200 MPa, the internal damage parameter $(\Delta W/W_e)$ decreases with the interface shear stress, corresponding to the interface slip case (Case 4) in Table 4.3. Under $\sigma_{max} = 200$ MPa, the broken fiber fraction (P) is higher than that under $\sigma_{max} = 170$ MPa at the same interface shear stress, and the Kachanov's damage parameter (D) is also higher than that under $\sigma_{max} = 170$ MPa at the same cycle number. However, the value of the interface shear stress under $\sigma_{max} = 200$ MPa is close to that under $\sigma_{max} = 170$ MPa.

Under $\sigma_{max} = 170$ MPa, the cycle-dependent internal friction parameter $(\Delta W/W_e)$ increases to the peak value first and then decreases, that is, from $\Delta W/W_e = 0.2$ at $\tau_i = 150$ MPa to $\Delta W/W_e = 0.42$ at $\tau_i = 39.5$ MPa, and then to $\Delta W/W_e = 0.04$ at $\tau_i = 1.25$ MPa. The broken fiber fraction (P) increases from $P = 0.0007$ at $\tau_i = 150$ MPa to $P = 0.12$ at $\tau_i = 1.2$ MPa. The Kachanov's damage parameter (D) increases from $D = 0$ at $N = 1$ to $D = 0.068$ at

TABLE 4.4   Cycle-Dependent Damage Evolution of 2.5D Woven Self-healing Hi-Nicalon™ SiC/[Si-B-C] Composite at $T = 1200°C$ in an Air Atmosphere

Cycle Number	$\Delta W/W_e$	τ_i/MPa	$P/\%$
0/170 MPa at 1200°C in an Air Atmosphere			
1	0.097	5.5	2.0
10	0.076	4.7	2.4
517	0.068	4.3	2.6
1281	0.062	3.6	3.2
13,700	0.047	2.5	4.9
32,334	0.044	1.7	7.8
−50/200 MPa at 1200°C in an Air Atmosphere			
1	0.133	5.1	2.7
5	0.096	3.6	4.0
20	0.090	3.4	4.3
100	0.087	3.3	4.4
450	0.085	3.2	4.6
790	0.082	3.1	4.8
8910	0.073	2.6	5.9
66,794	0.061	2.0	8.1

$N = 13,202$. The interface shear stress (τ_i) decreases from $\tau_i = 5.5$ MPa at $N = 1$ to $\tau_i = 1.7$ MPa at $N = 32334$.

Under $\sigma_{max} = 200$ MPa, the cycle-dependent internal friction parameter ($\Delta W/W_e$) increases to the peak value first and then decreases, that is, from $\Delta W/W_e = 0.161$ at $\tau_i = 150$ MPa to $\Delta W/W_e = 0.42$ at $\tau_i = 26.4$ MPa, and then to $\Delta W/W_e = 0.05$ at $\tau_i = 1.37$ MPa. The broken fiber fraction (P) increases from $P = 0.0008$ at $\tau_i = 150$ MPa to $P = 0.14$ at $\tau_i = 1.3$ MPa. The Kachanov's damage parameter (D) increases from $D = 0$ at $N = 1$ to $D = 0.2$ at $N = 60530$. The interface shear stress (τ_i) decreases from $\tau_i = 5.1$ MPa at $N = 1$ to $\tau_i = 2$ MPa at $N = 66794$.

4.5 SUMMARY AND CONCLUSION

In this chapter, cycle-dependent damage evolution of self-healing 2.5D woven Hi-Nicalon™ SiC/[Si-B-C] composite under different peak stresses at $T = 600$ and 1200 °C was investigated. The damage parameters of internal friction ($\Delta W/W_e$), dissipated energy (ΔW), Kachanov's damage parameter (D), broken fiber fraction (P), and interface shear stress (τ_i) were used to describe fatigue damage evolution. For 2.5D woven self-healing

Hi-Nicalon™ SiC/[Si-B-C] composite, temperature is a governing parameter for the fatigue process. At T = 600°C in an air atmosphere, $\Delta W/W_e$ first decreases and then increases with cycle number, and at T = 1200°C in an air atmosphere, $\Delta W/W_e$ decreases with cycle number. The degradation rate of the interface shear stress and broken fiber faction increases with peak stress.

REFERENCES

1. Lamouroux F, Camus G, Thebault J. Kinetics and mechanisms of oxidation of 2D woven C/SiC composites: I, Experimental approach. *J. Am. Ceram. Soc.* 1994; 77:2049–2057.
2. Hay RS, Chater RJ. Oxidation kinetics strength of Hi-Nicalon™-S SiC fiber after oxidation in dry and wet air. *J. Am. Ceram. Soc.* 2017; 100:4110–4130.
3. Viricelle JP, Goursat P, Bahloul-Hourlier D. Oxidation behavior of a multi-layered ceramic-matrix composite $(SiC)_f/C/(SiBC)_m$. *Compos. Sci. Technol.* 2001; 61:607–614.
4. Bouillon E, Abbe F, Goujard S, Pestourie E, Habarou G. Mechanical and thermal properties of a self-sealing matrix composite and determination of the life time duration. *Ceram. Eng. Sci. Proc.* 2002; 21:459–467.
5. Ruggles-Wrenn MB, Delapasse J, Chamberlain AL, Lane JE, Cook TS. Fatigue behavior of a Hi-Nicalon™/SiC-B_4C composite at 1200°C in air and in steam. *Mater. Sci. Eng. A* 2012; 534:119–128.
6. Ruggles-Wrenn MB, Kurtz G. Notch sensitivity of fatigue behavior of a Hi-Nicalon™/SiC-B_4C composite at 1200°C in air and in steam. *Appl. Compos. Mater.* 2013; 20:891–905.
7. Ruggles-Wrenn MB, Lee MD. Fatigue behavior of an advanced SiC/SiC composite with a self-healing matrix at 1300°C in air and in steam. *Mater. Sci. Eng. A* 2016; 677:438–445.
8. Reynaud P, Rouby D, Fantozzi G. Cyclic fatigue behaviour at high temperature of a self-healing ceramic matrix composite. *Ann. Chim. Sci. Mater.* 2005; 30:649–658.
9. Carrere P, Lamon J. Fatigue behavior at high temperature in air of a 2D woven SiC/SiBC with a self healing matrix. *Key Eng. Mater.* 1999; 164–165:321–324.
10. Forio P, Lamon J. Fatigue behavior at high temperatures in air of a 2D SiC/Si-B-C composite with a self-healing multilayered matrix. *Adv. Ceram. Matrix Compos. VII* 2001.
11. Simon C, Rebillat F, Camus G. Electrical resistivity monitoring of a SiC/[Si-B-C] composite under oxidizing environments. *Acta Mater.* 2017; 132:586–597.
12. Simon C, Rebillat F, Herb V, Camus G. Monitoring damage evolution of SiC_f/[Si-B-C]$_m$ composites using electrical resistivity: Crack density-based electromechanical modeling. *Acta Mater.* 2017; 124:579–587.

13. Moevus M, Reynaud P, R'Mili M, Godin N, Rouby D, Fantozzi G. Static fatigue of a 2.5D SiC/[Si-B-C] composite at intermediate temperature under air. *Adv. Sci. Technol.* 2006; 50:141–146.

14. Reynaud P. Cyclic fatigue of ceramic-matrix composites at ambient and elevated temperatures. *Compos. Sci. Technol.* 1996; 56:809–814.

15. Dalmaz A, Reynaud P, Rouby D, Fantozzi G, Abbe F. Mechanical behavior and damage development during cyclic fatigue at high-temperature of a 2.5D carbon/sic composite. *Compos. Sci. Technol.* 1998; 58:693–699.

16. Fantozzi G, Reynaud P. Mechanical hysteresis in ceramic matrix composites. *Mater. Sci. Eng. Part A* 2009; 521–522:18–23.

17. Li LB. A hysteresis dissipated energy-based damage parameter for life prediction of carbon fiber-reinforced ceramic-matrix composites under fatigue loading. *Compos. Part B Eng.* 2015; 82:108–128.

18. Li LB. Damage monitoring and life prediction of carbon fiber-reinforced ceramic-matrix composites at room and elevated temperatures using hysteresis dissipated energy-based damage parameter. *Compos. Interface* 2018; 25:335–356.

19. Li LB, Reynaud P, Fantozzi G. Mechanical hysteresis and damage evolution in C/SiC composites under fatigue loading at room and elevated temperatures. *Int. J. Appl. Ceram. Technol.* 2019; 16:2214–2228.

20. Li LB. Failure analysis of long-fiber-reinforced ceramic-matrix composites subjected to in-phase thermomechanical and isothermal cyclic loading. *Eng. Fail. Anal.* 2019; 104:856–872.

21. Li LB, Reynaud P, Fantozzi G. Cyclic-dependent damage evolution in self-healing woven SiC/[Si-B-C] ceramic-matrix composites at elevated temperatures. *Materials* 2020; 13:1478.

22. Penas O. Etude de Composites SiC/SiBC à Matrice Multiséquencée en Fatigue Cyclique à Hautes Températures Sous Air. Ph.D. Thesis, INSA de Lyon, Villeurbanne, France, 2002.

23. Zhu SJ, Mizuno M, Kagawa Y, Mutoh Y. Monotonic tension, fatigue and creep behavior of SiC-fiber-reinforced SiC-matrix composites: A review. *Compos. Sci. Technol.* 1999, 59:833–851,

24. Ruggles-Wrenn MB, Christensen DT, Chamberlain AL, Lane JE, Cook TS. Effect of frequency and environment on fatigue behavior of a CVI SiC/SiC ceramic matrix composite at 1200°C. *Compos. Sci. Technol.* 2011; 71:190–196.

25. Ruggles-Wrenn MB, Jones TP. Tension-tension fatigue of a SiC/SiC ceramic matrix composite at 1200°C in air and in steam. *Int. J. Fatigue* 2013; 47:154–160.

26. Li LB. Damage and failure of fiber-reinforced ceramic-matrix composites subjected to cyclic fatigue, dwell fatigue and thermomechanical fatigue. *Ceram. Int.* 2017; 43:13978–13996.

27. Li LB. Time-dependent damage and fracture of fiber-reinforced ceramic-matrix composites at elevated temperatures. *Compos. Interfaces* 2019; 26:963–988.

CHAPTER 5

High-Temperature Static-Fatigue Mechanical Hysteresis Behavior in Two-Dimensional Plain-Woven Chemical Vapor Infiltration C/[Si-B-C] Composites

5.1 INTRODUCTION

Ceramic materials possess high strength and modulus at elevated temperatures. But their use as structural components is severely limited because of their brittleness. Continuous fiber-reinforced ceramic-matrix composites (CMCs), by incorporating fibers in ceramic matrices, however, not only exploit their attractive high-temperature strength but also reduce the propensity for catastrophic failure. Carbon fiber–reinforced silicon carbide ceramic-matrix composites (C/SiC CMCs) are one of the most promising candidates for many high-temperature applications, particularly as aerospace and aircraft thermostructural components [1–3]. However, one

DOI: 10.1201/b23026-5

of the barriers to their use in certain long-term or reusable applications is that degradation of the carbon fibers in oxidizing environments can lead to strength reduction and component failure [4].

Many researchers have performed experimental and theoretical investigations on the effects of oxidation damage on the mechanical behavior of fiber-reinforced CMCs. In the experimental research area, Zhu [5] investigated the effect of oxidation on the fatigue behavior of two-dimensional (2D) SiC/SiC composite at elevated temperatures. It was found that the fatigue life decreased 13% after oxidation at $T = 600°C$ for $t = 100$ hours due to the disappearance of carbon interphase. Mall and Engesser [6] investigated the damage evolution in 2D C/SiC composite under different fatigue loading frequencies at $T = 550°C$ in an air atmosphere. The oxidation of carbon fibers caused a reduction in fatigue life of C/SiC composite under lower loading frequency. However, the oxidation of carbon fibers was almost absent or negligible at higher frequency at elevated temperatures. Fantozzi and Reynaud [7] investigated the static-fatigue behavior of C/[Si-B-C] composite at $T = 1200°C$ in air atmosphere. The areas of stress-strain hysteresis loops after a static fatigue of $t = 144$ hours have significantly decreased, attributed to time dependent of fiber/matrix PyC interface recession by oxidation or by a beginning of carbon fibers recession by oxidation. In the theoretical research area, much work has been conducted to analyze and model the oxidation of fibers, matrices and interfaces without loading by assuming steady-state diffusion of oxidation [8, 9]. Halbig et al. [10] investigated the stressed-oxidation of different fiber-reinforced CMCs, that is, C/SiC, SiC/SiC, and SiC/SiNC, among others, and developed a model to predict the oxidation pattern and kinetics of carbon fiber tows in a nonreactive matrix. Pailler and Lamon [11] developed a fatigue-oxidation model to investigate the strain response of a SiC/SiC minicomposite under matrix cracking and interface oxidation. Casas et al. [12] developed a creep-oxidation model for fiber-reinforced CMCs at elevated temperatures, including the effects of interface and matrix oxidation, the creep of fibers, and the degradation of fibers strength with time. The broken fibers fraction increases with time in an accelerated manner due to fibers strength degradation. Under static-fatigue loading at elevated temperature, the shape, location, and area of the stress-strain hysteresis loops would evolve with an increase in the oxidation time, which can be used to monitor the damage evolution inside of the damaged composite [13].

The objective of this chapter is to develop the hysteresis loops models of fiber-reinforced CMCs considering interface oxidation at elevated temperature. The oxidation region propagating model is adopted to analyze the oxidation effect on the hysteresis loops of the composite under static-fatigue at elevated temperature, which is controlled by interface frictional slip between the fiber and the matrix, and diffusion of oxygen gas through matrix cracks. Based on the damage mechanism of fiber sliding relative to matrix in the interface debonded region upon unloading and subsequent reloading, the hysteresis loops models corresponding to different interface slip cases considering interface oxidation are established. Relationships between the hysteresis loops, hysteresis dissipated energy, interface frictional slip, and oxidation duration are established. Experimental hysteresis loops of C/[Si-B-C] composite under static fatigue in air at $T = 1200$ °C are predicted. Effects of stress level, matrix crack spacing, fiber volume content, and oxidation temperature on the static-fatigue hysteresis loops are analyzed.

5.2 MICROMECHANICAL HYSTERESIS CONSTITUTIVE MODEL

If matrix cracking and fiber/matrix interface debonding are present on the first loading to peak stress, the stress-strain hysteresis loops develop as a result of energy dissipation through frictional slip between fibers and matrix upon unloading/reloading. At elevated temperatures, matrix cracks will serve as avenues for the ingress of the environment atmosphere into the composite. The oxygen reacts with carbon layer along the fiber length at a certain rate of $d\xi/dt$, where ξ is the length of carbon lost in each side of the crack.

$$\xi = \varphi_1 \left[1 - \exp\left(-\frac{\varphi_2 t}{b} \right) \right], \qquad (5.1)$$

where φ_1 and φ_2 are parameters dependent on temperature and described using the Arrhenius type laws [12],

$$\varphi_1 = 7.021 \times 10^{-3} \times \exp\left(\frac{8231}{T} \right), \qquad (5.2)$$

$$\varphi_2 = 227.1 \times \exp\left(-\frac{17090}{T} \right), \qquad (5.3)$$

where φ_1 is in mm and φ_2 in s^{-1}; φ_1 represents the asymptotic behavior for long times, which decreases with temperature; the product $\varphi_1\varphi_2$ represents the initial oxidation rate, which is an increasing function of temperature; T is the absolute temperature; and b is a delay factor considering the deceleration of reduced oxygen activity.

The interface shear stress in the oxidized region decreases from the initial value τ_i to τ_f due to interface oxidation. With the increase of oxidation duration under static fatigue, the oxidized region propagates along the fiber/matrix interface, leading to the evolution of the shape, area, and location of the mechanical hysteresis loops.

Based on the interface frictional slip cases between fibers and matrix, the stress-strain hysteresis loops under static fatigue at elevated temperature can be divided into four different cases:

- Case 1, the interface partial debonding and the fiber complete sliding relative to the matrix

- Case 2, the interface partial debonding and the fiber partial sliding relative to the matrix

- Case 3, the interface complete debonding and the fiber partial sliding relative to matrix

- Case 4, the interface complete debonding and the fiber complete sliding relative to the matrix in the interface debonding region

The unloading and reloading stress-strain relationships for interface partial debonding and fiber partial sliding relative to the matrix are

$$
\begin{aligned}
\varepsilon_{c\_pu} = {} & \frac{2\sigma L_d}{V_f E_f L_c} + \frac{2\tau_f}{r_f E_f L_c}\xi^2 + \frac{4\tau_f}{r_f E_f L_c}\xi\left(L_d - \xi\right) \\
& + \frac{4\tau_i}{r_f E_f L_c}\left(L_y - \xi\right)^2 - \frac{2\tau_i}{r_f E_f L_c}\left(2L_y - \xi - L_d\right)^2 + \frac{2\sigma_{fo}}{E_f L_c}\left(\frac{L_c}{2} - L_d\right) \\
& + \frac{2r_f}{\rho E_f L_c}\left[\frac{V_m}{V_f}\sigma_{mo} + \frac{2\tau_f}{r_f}\xi + \frac{2\tau_i}{r_f}\left(2L_y - \xi - L_d\right)\right] \\
& \times \left[1 - \exp\left(-\rho\frac{L_c/2 - L_d}{r_f}\right)\right] - \left(\alpha_c - \alpha_f\right)\Delta T,
\end{aligned}
\tag{5.4}
$$

$$\varepsilon_{c\_pr} = \frac{2\sigma}{V_f E_f L_c} L_d - \frac{4\tau_f}{r_f E_f L_c} L_z^2 + \frac{2\tau_f}{r_f E_f L_c}\left(2L_z - \xi\right)^2$$

$$- \frac{4\tau_f}{r_f E_f L_c}\left(2L_z - \xi\right)\left(L_d - \xi\right) + \frac{4\tau_i}{r_f E_f L_c}\left(L_y - \xi\right)^2$$

$$- \frac{2\tau_i}{r_f E_f L_c}\left(2L_y - \xi - L_d\right)^2 + \frac{2\sigma_{fo}}{E_f L_c}\left(\frac{L_c}{2} - L_d\right)$$

$$+ \frac{2r_f}{\rho E_f L_c}\left[\frac{V_m}{V_f}\sigma_{mo} - \frac{2\tau_f}{r_f}\left(2L_z - \xi\right) + \frac{2\tau_i}{r_f}\left(2L_y - \xi - L_d\right)\right]$$

$$\times \left[1 - \exp\left(-\rho\frac{L_c/2 - L_d}{r_f}\right)\right] - \left(\alpha_c - \alpha_f\right)\Delta T, \tag{5.5}$$

where V_f is the fiber volume fraction; E_f is the elastic modulus of the fiber; r_f is the fiber radius; L_d is the interface debonding length; L_c is the matrix crack spacing; L_y and L_z are the interface counter-slip length and interface new-slip length, τ_f and τ_i are the interface shear stress in the oxidation and slip region, respectively; α_f, and α_c denote the fiber and the composite thermal expansion coefficient, respectively; and ΔT denotes the temperature difference between the fabricated temperature T_0 and room temperature T_1 ($\Delta T = T_1 - T_0$).

When the fiber complete slides relative to the matrix, the unloading stress-strain relationship can be divided into two regions, that is, (1) when $\sigma > \sigma_{tr\_pu}$, the unloading strain is determined by Equation 5.4, and (2) when $\sigma < \sigma_{tr\_pu}$, the unloading strain is determined by Equation 5.4 by setting $L_y = L_d$. The reloading stress-strain relationship can be divided into two regions, that is, (1) when $\sigma < \sigma_{tr\_pr}$, the reloading strain is determined by Equation 5.5 and (2) when $\sigma > \sigma_{tr\_pr}$, the reloading strain is determined by Equation 5.5 by setting $L_z = L_d$.

The unloading and reloading stress-strain relationships for interface complete debonding and fiber partial sliding relative to the matrix are

$$\varepsilon_{c\_fu} = \frac{\sigma}{V_f E_f} - \frac{2\tau_f}{r_f E_f L_c}\xi^2 + \frac{2\tau_f}{r_f E_f}\xi + \frac{4\tau_i}{r_f E_f L_c}\left(L_y - \xi\right)^2$$

$$- \frac{2\tau_i}{r_f E_f L_c}\left(2L_y - \xi - \frac{L_c}{2}\right)^2 - \left(\alpha_c - \alpha_f\right)\Delta T, \tag{5.6}$$

$$\varepsilon_{c\_fr} = \frac{\sigma}{V_f E_f} - \frac{4\tau_f}{r_f E_f L_c} L_z^2 + \frac{2\tau_f}{r_f E_f L_c}\left(2L_z - \xi\right)^2$$

$$- \frac{4\tau_f}{r_f E_f L_c}\left(2L_z - \xi\right)\left(L_y - \xi\right) + \frac{4\tau_i}{r_f E_f L_c}\left(L_y - \xi\right)^2$$

$$- \frac{4\tau_f}{r_f E_f L_c}\left(2L_z - \xi\right)\left(\frac{L_c}{2} - L_y\right) \tag{5.7}$$

$$- \frac{2\tau_i}{r_f E_f L_c}\left(2L_y - \xi - \frac{L_c}{2}\right)^2 - \left(\alpha_c - \alpha_f\right)\Delta T.$$

When the fiber complete slides relative to the matrix, the unloading stress-strain relationship can be divided into two regions, that is, (1) when $\sigma > \sigma_{tr\_fu}$, the unloading strain is determined by Equation 5.6, and (2) when $\sigma < \sigma_{tr\_fu}$, the unloading strain is determined by Equation 5.6 by setting $L_y = L_c/2$. The reloading stress-strain relationship can be divided into two parts, that is, (1) when $\sigma < \sigma_{tr\_fr}$, the reloading strain is determined by Equation 5.7, and (2) when $\sigma > \sigma_{tr\_fr}$, the reloading strain is determined by Equation 5.7 by setting $L_z = L_c/2$.

Under cyclic loading, the area associated with the stress-strain hysteresis loops is the energy lost during the corresponding cycle, which is defined as

$$\Delta W = \int_{\sigma_{min}}^{\sigma_{max}} \left[\varepsilon_c^{unload}\left(\sigma\right) - \varepsilon_c^{reload}\left(\sigma\right)\right] d\sigma. \tag{5.8}$$

5.3 EXPERIMENTAL COMPARISONS

Fantozzi and Reynaud [7] investigated the stress-strain hysteresis loops of C/[Si-B-C] composite during static fatigue at $T = 1200°C$ in air and an applied stress $\sigma = 170$ MPa. The load applied on the composite is steady and periodically performed an unloading/reloading sequence to obtain the stress-strain hysteresis loops.

The experimental and predicted stress-strain hysteresis loops corresponding to different oxidation duration, that is, from first loading to $t = 144$ hours static fatigue in air, are shown in Figure 5.1a. The hysteresis

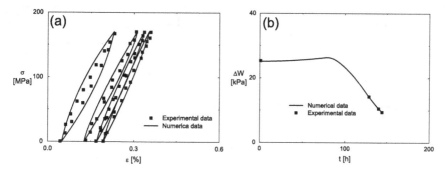

FIGURE 5.1 (a) Experimental and predicted stress-strain hysteresis loops corresponding to different oxidation duration and (b) experimental and predicted hysteresis dissipated energy versus oxidation duration of C/[Si-B-C] composite under static fatigue in air at $T = 1200°C$ and an applied stress of $\sigma_{max} = 170$ MPa.

dissipated energy versus oxidation duration curve is shown in Figure 5.1b, in which the hysteresis dissipated energy increases from $\Delta W = 25.3$ kPa to the peak value $\Delta W = 26.3$ kPa first and then decreases with increasing oxidation duration to $\Delta W = 9.4$ kPa at 144 hours, attributed to the propagation of oxidation region with decreasing interface shear stress in the oxidized region.

The static fatigue hysteresis loops correspond to the interface slip Case 1 on first loading, that is, the interface partial debonding and the fiber partial sliding relative to the matrix in the interface debonding region; then with increasing oxidation duration, the hysteresis loops correspond to the interface slip Case 4, that is, the interface completely debonds and the fiber completely slides relative to the matrix in the interface debonding region.

The predicted hysteresis loops and the evolution of hysteresis dissipated energy versus oxidation duration agreed with experimental data.

5.4 DISCUSSION

5.4.1 Effect of Stress Level on Static Fatigue Hysteresis Behavior

The hysteresis dissipated energy, interface debonding, oxidation, and slip lengths versus oxidation duration under $\sigma_{max} = 180$ and 200 MPa are shown in Figure 5.2.

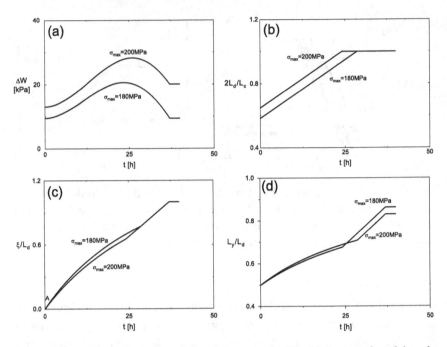

FIGURE 5.2   (a) The hysteresis dissipated energy (ΔW), (b) the interface debonding ratio ($2L_d/L_c$), (c) the interface oxidation ratio (ξ/l_d), and (d) the interface counter-slip ratio (L_y/L_d) under σ_{max} = 180 and 200 MPa.

Under σ_{max} = 180 MPa, the hysteresis dissipated energy versus oxidation duration curve can be divided into three regions:

(1) In region I, the hysteresis dissipated energy increases from ΔW = 9.5 kPa to the peak value ΔW = 20.5 kPa and decreases to ΔW = 18.9 kPa. The interface partially debonds and partially oxidizes, and the fiber partially slides relative to the matrix.

(2) In region II, the hysteresis dissipated energy continually decreases to ΔW = 9.3 kPa. The interface completely debonds and partially oxidizes, and the fiber partially slides relative to the matrix.

(3) In region III, the hysteresis dissipated energy remains to be constant value ΔW = 9.3 kPa. The interface completely debonds and completely oxidizes, and the fiber partially slides relative to the matrix.

Under σ_{max} = 200 MPa, the hysteresis dissipated energy versus oxidation duration curve can be divided into three regions:

(1) In region I, the hysteresis dissipated energy increases from ΔW = 13 kPa to the peak value ΔW = 28.2 kPa and decreases to ΔW = 28 kPa. The interface partially debonds and partially oxidizes, and the fiber partially slides relative to the matrix.

(2) In region II, the hysteresis dissipated energy decreases to ΔW = 20 kPa. The interface completely debonds and partially oxidizes, and the fiber partially slides relative to the matrix.

(3) In region III, the hysteresis dissipated energy remains to be constant value ΔW = 20 kPa. The interface completely debonds and completely oxidizes, and the fiber partially slides relative to the matrix.

When the static peak stress increases, the hysteresis dissipated energy increases for different oxidation durations, and the oxidation duration corresponding to complete interface debonding decreases.

5.4.2 Effect of Matrix Crack Spacing on Static-Fatigue Hysteresis Behavior

The hysteresis dissipated energy, interface debonding, oxidation, and slip lengths versus oxidation duration for L_c = 200 and 240 μm are shown in Figure 5.3.

When L_c = 200 μm, the hysteresis dissipated energy versus oxidation duration curve is divided into three regions:

(1) In region I, the hysteresis dissipated energy increases from ΔW = 10.4 kPa to the peak value ΔW = 22.6 kPa and decreases to ΔW = 22.5 kPa. The interface partially debonds and partially oxidizes, and the fiber partially slides relative to the matrix.

(2) In region II, the hysteresis dissipated energy continually decreases to ΔW = 15.6 kPa. The interface completely debonds and partially oxidizes, and the fiber partially slides relative to the matrix.

(3) In region III, the hysteresis dissipated energy remains to be constant value ΔW = 15.6 kPa. The interface completely debonds and completely oxidizes, and the fiber partially slides relative to the matrix.

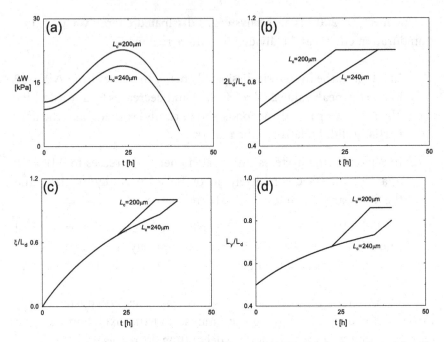

FIGURE 5.3 (a) The hysteresis dissipated energy (ΔW), (b) the interface debonding ratio ($2L_d/L_c$), (c) the interface oxidation ratio (ξ/l_d), and (d) the interface counter-slip ratio (L_y/L_d) for $L_c = 200$ and 240 μm.

When $l_c = 240$ μm, the hysteresis dissipated energy versus oxidation duration curve is divided into two regions:

(1) In region I, the hysteresis dissipated energy increases from $\Delta W = 8.7$ kPa to the peak value $\Delta W = 18.8$ kPa and decreases to $\Delta W = 11.5$ kPa. The interface partially debonds and partially oxidizes, and the fiber partially slides relative to the matrix.

(2) In region II, the hysteresis dissipated energy continually decreases to $\Delta W = 3.6$ kPa. The interface completely debonds and partially oxidizes, and the fiber partially slides relative to the matrix.

When the matrix crack spacing increases, the hysteresis dissipated energy decreases for different oxidation duration, and the oxidation duration for interface complete debonding increases.

5.4.3 Effect of Fiber's Volume Fraction on Static-Fatigue Hysteresis Behavior

The hysteresis dissipated energy, interface debonding, oxidation, and slip lengths versus oxidation duration for $V_f = 35\%$ and 40% are shown in Figure 5.4.

When $V_f = 35\%$, the hysteresis dissipated energy versus oxidation duration curve is divided into three regions:

(1) In region I, the hysteresis dissipated energy increases from $\Delta W = 15.9$ to 29.4 kPa. The interface partially debonds and partially oxidizes, and the fiber partially slides relative to the matrix.

(2) In region II, the hysteresis dissipated energy continually increases to the peak value $\Delta W = 34.5$ kPa. The interface completely debonds and partially oxidizes, and the fiber partially slides relative to the matrix.

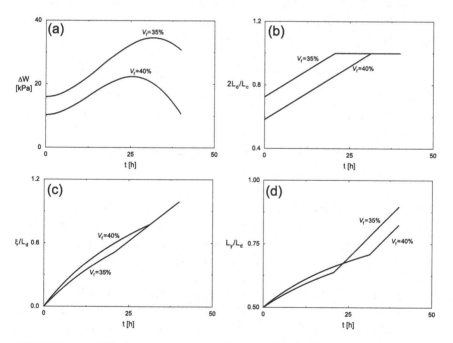

FIGURE 5.4   (a) The hysteresis dissipated energy (ΔW), (b) the interface debonding ratio ($2L_d/L_c$), (c) the interface oxidation ratio (ξ/l_d), and (d) the interface counter-slip ratio (L_y/L_d) for $V_f = 35$ and 40%.

(3) In region III, the hysteresis dissipated energy decreases to $\Delta W = 30.7$ kPa. The interface completely debonds and partially oxidizes, and the fiber partially slides relative to the matrix.

When $V_f = 40\%$, the hysteresis dissipated energy versus oxidation duration curve is divided into two regions:

(1) In region I, the hysteresis dissipated energy increases from $\Delta W = 10.3$ kPa to the peak value $\Delta W = 22.3$ kPa and decreases to $\Delta W = 20.4$ kPa. The interface partially debonds and partially oxidizes, and the fiber partially slides relative to the matrix.

(2) In region II, the hysteresis dissipated energy continually decreases to $\Delta W = 10.6$ kPa. The interface completely debonds and partially oxidizes, and the fiber partially slides relative to the matrix.

When the fiber volume content increases, the hysteresis dissipated energy decreases for different oxidation duration, and the oxidation duration for the complete interface debonding increases.

5.4.4 Effect of Temperature on Static-Fatigue Hysteresis Behavior

The hysteresis dissipated energy, interface debonding, oxidation, and slip lengths versus oxidation duration for $T = 800$ and 900°C are shown in Figure 5.5.

When $T = 800°C$, the hysteresis dissipated energy versus oxidation duration curve is divided into three regions:

(1) In region I, the hysteresis dissipated energy increases from $\Delta W = 9.5$ kPa to the peak value $\Delta W = 20.5$ kPa and decreases to $\Delta W = 18.9$ kPa. The interface partially debonds and partially oxidizes, and the fiber partially slides relative to the matrix.

(2) In region II, the hysteresis dissipated energy continually decreases to $\Delta W = 9.3$ kPa. The interface completely debonds and partially oxidizes, and the fiber partially slides relative to the matrix.

(3) In region III, the hysteresis dissipated energy remains to be constant value $\Delta W = 9.3$ kPa. The interface completely debonds and oxidizes, and the fiber partially slides relative to the matrix.

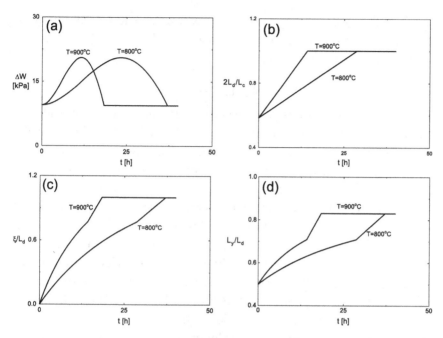

FIGURE 5.5 (a) The hysteresis dissipated energy (ΔW), (b) the interface debonding ratio ($2L_d/L_c$), (c) the interface oxidation ratio (ξ/l_d), and (d) the interface counter-slip ratio (L_y/L_d) for T = 800 and 900°C.

When T = 900°C, the hysteresis dissipated energy versus oxidation duration curve is divided into three regions:

(1) In region I, the hysteresis dissipated energy increases from ΔW = 9.5 kPa to the peak value ΔW = 20.5 kPa and decreases to ΔW = 18.9 kPa. The interface partially debonds and oxidizes, and the fiber partially slides relative to the matrix.

(2) In region II, the hysteresis dissipated energy continually decreases to ΔW = 9.3 kPa. The interface completely debonds and partially oxidizes, and the fiber partially slides relative to the matrix.

(3) In region III, the hysteresis dissipated energy remains to be constant value ΔW = 9.3 kPa. The interface completely debonds and oxidizes, and the fiber partially slides relative to the matrix.

When the oxidation temperature increases, the oxidation duration for the complete interface debonding decreases, and the peak hysteresis dissipated energy remains constant.

5.5 SUMMARY AND CONCLUSION

An analytical method was developed to investigate the effect of oxidation on the hysteresis loops of fiber-reinforced CMCs under static fatigue at elevated temperatures. The oxidation region propagating model was adopted to analyze the oxidation effect on the hysteresis loops of the composite under static fatigue at elevated temperatures, which is controlled by interface frictional slip between the fiber and the matrix and the diffusion of oxygen gas through matrix crackings. Based on the damage mechanism of fiber sliding relative to the matrix in the interface debonding region on unloading and subsequent reloading, the hysteresis loops models corresponding to different interface slip cases considering interface oxidation were established. Relationships between the hysteresis loops, hysteresis dissipated energy, and interface frictional slip and oxidation duration were established. Experimental hysteresis loops of C/[Si-B-C] composite under static fatigue in air at 1200°C were predicted. Effects of stress level, matrix crack spacing, fiber volume content, and oxidation temperature on the hysteresis dissipated energy, interface debonding, and oxidation and frictional slip lengths versus oxidation duration were analyzed. The hysteresis dissipated energy increases to the peak value first, then decreases with the increase of oxidation duration, attributed to the propagation of oxidation region with the decrease of interface shear stress in the oxidized region.

REFERENCES

1. Naslain R. Design, preparation and properties of non-oxide CMCs for application in engines and nuclear reactors: an overview. *Compos. Sci. Technol.* 2004; 64:155–170.
2. Santhosh U, Ahmad J, John R, Ojard G, Miller R, Gowayed Y. Modeling of stress concentration in ceramic matrix composites. *Compos. Part B: Eng.* 2013; 45:1156–1163.
3. Murthy Pappu LN, Nemeth NN, Brewer DN, Mital S. Probabilistic analysis of a SiC/SiC ceramic matrix composite turbine vane. *Compos. Part B: Eng.* 2008; 39:694–703.
4. Naslain R, Guette A, Rebillat F, Gallet S, Lamouroux F, Filipuzzi L, Louchet C. Oxidation mechanisms and kinetics of SiC-matrix composites and their constituents. *J. Mater. Sci.* 2004; 39:7303–7316.

5. Zhu SJ. Fatigue behavior of ceramic matrix composite oxidized at intermediate temperature. *Mater. Trans.* 2006; 47:1965–1967.
6. Mall S, Engesser JM. Effects of frequency on fatigue behavior of CVI C/SiC at elevated temperature. *Compos. Sci. Technol.* 2006; 66:863–874.
7. Fantozzi G, Reynaud P. Mechanical hysteresis in ceramic matrix composites. *Mater. Sci. Eng. A* 2009; 521–522:18–23.
8. Lamouroux F, Naslain R, Jouin JM. Kinetics and mechanisms of oxidation of 2D woven C/SiC composites: II, theoretical approach. *J. Am. Ceram. Soc.* 1994; 77:2058–2068.
9. Sullivan RM. A model for the oxidation of carbon silicon carbide composite structures. *Carbon* 2005; 43:275–285.
10. Halbig MC, Brewer DN, Eckel AJ. Degradation of continuous fiber ceramic matrix composites under constant-load conditions. NASA/TM-2000-209681.
11. Pailler F, Lamon J. Micromechanics based model of fatigue/oxidation for ceramic matrix composites. *Compos. Sci. Technol.* 2005; 65:369–374.
12. Casas L, Martinez-Esnaola JM. Modelling the effect of oxidation on the creep behavior of fiber-reinforced ceramic matrix composites. *Acta Mater.* 2003; 51:3745–3757.
13. Li LB. Modeling the effect of oxidation on hysteresis loops of carbon fiber-reinforced ceramic-matrix composites under static fatigue at elevated temperature. *J. Eur. Ceram. Soc.* 2016; 36:465–480.

CHAPTER 6

High-Temperature Dwell-Fatigue Mechanical Hysteresis Behavior in Cross-Ply SiC/MAS Composites

6.1 INTRODUCTION

Ceramic-matrix composites (CMCs) possess high strength-to-weight ratios at elevated temperatures and are being designed and developed for hot section components in commercial aero engine [1–3]. As new materials, the CMCs need to meet the airworthiness certification requirements [4, 5], and it is necessary to analyze the degradation, damage, and failure mechanisms subjected to cyclic loading at different temperatures and environments [6, 7].

Many researchers performed experimental and theoretical investigations on the dwell-fatigue behavior of fiber-reinforced CMCs. Zhu [8] investigated the monotonic tension, creep, and tension-tension cyclic fatigue behavior of 2D Nicalon™ SiC/SiC composite at 1300°C in an air and argon environment. The slope decreases and the width of the loops increases with applied cycles. The hysteresis loops move to the right along the strain axis due to the time-dependent damage, that is, matrix cracking, fiber/matrix interface debonding, and oxidation. However, the relationships between the macro–stress/strain behavior and microstructure damages inside of SiC/SiC composite have not been established. John et al. [9] investigated the durability of melt infiltrated (MI) two-dimensional (2D) Hi-Nicalon™ SiC/SiC composite for different loading frequencies, that is, f = 1 and 30 Hz, and dwell fatigue and creep loading at T = 815 and 1204 °C, respectively. It was found that the periodic loading and unloading cycles degrade the material performance of the dwell-fatigue specimens compared with that of creep specimens. The lifetime under fatigue and creep for different peak stresses were compared and analyzed; however, the micro-damage mechanisms that caused the lifetime difference have not been analyzed. Gowayed et al. [10] investigated the accumulation of time-dependent strain under dwell fatigue of MI 2D Sylramic™ SiC/SiC composites with and without holes at T = 815 and 1204°C, respectively. The time-dependent strain accumulation was observed at the maximum stress level caused by matrix crack opening and oxidation of the boron nitride (BN) coating. Upon unloading and reloading, the specimen breaks the seal and allows further ingress of the environment, which degrades the material performance of the composite under dwell-fatigue loading. However, the accumulation of the composite time-dependent strain was predicted by data fitting using a 3-parameter curve fitting approach, not the damage-based models or approaches. Ruggles-Wrenn and Sharma [11] investigated the cyclic tension-tension fatigue behavior of 2D Sylramic™ SiC/[SiC+Si_3N_4] composite at 1300°C in air and in steam atmospheres. At higher fatigue peak stress, the presence of steam caused noticeable degradation in the fatigue performance of the composite. The evolution of the peak strain and normalized modulus versus cycle fatigue cycles have been analyzed. Ruggles-Wrenn et al. [12, 13] investigated the cyclic tension-tension fatigue behavior of 2D Hi-Nicalon™ SiC/SiC composite with an inhibited matrix at 1200 and 1300°C in air and in steam conditions, respectively. The presence of steam has little influence on the fatigue performance of the composite at f = 1.0 Hz but noticeably degrades fatigue lifetimes at f = 0.1 Hz. The flaw growth mechanisms in SiC fibers controlled the fatigue performance and lifetime of the SiC/SiC composite, which is affected by the testing temperature and environments. Shi et al. [14] investigated the low cycle tension-tension fatigue behavior of 3D KD-I™ SiC/SiC composite at T = 1300 °C in air condition. The effects of coating and loading frequency on the fatigue performance at elevated temperatures were investigated. The strain ratcheting and modulus degradation with applied cycles were correlated with damage evolution, that is, matrix cracking, interface debonding, and oxidation. However, the quantitative relationships between the damage evolution and the loading frequency, fatigue peak stress, and testing temperature have not been established. Maillet et al. [15] developed an acoustic emission (AE)–based approach to monitoring the damage evolution and lifetime of 2D SiC/[Si-B-C] composite with the self-healing matrix under static fatigue loading at 450 and 500°C. Singh [16] developed a novel approach to monitoring the crack initiation and growth in interlaminar testing of CMCs using the AE and direct current potential drop method. The damage modes of matrix crack, fibers breakage, and delamination can be detected. Zhang et al. [17] investigated the deformation histories and identified the locations of strain concentration for both un-notched and single-edge-notched oxide/oxide CMCs using the digital image correlation (DIC) technique. However, the acoustic emission and digital image correlation approaches have limitations on monitoring the multiple-stage damage evolution process under dwell fatigue, thermomechanical fatigue, or creep loading. Li [18, 19] established the relationship between the fatigue hysteresis dissipated energy–based damage parameter, internal damage of matrix cracking, fiber/matrix interface debonding and fiber fracture and predicted the damage evolution of unidirectional C/SiC under cyclic tension–tension fatigue loading at $T = 800$ °C in air condition [20]. Li [21, 22] investigated fatigue damage and lifetime of 2D SiC/SiC composite at elevated temperature without a hold time in air and in steam atmospheres, predicted the fatigue lifetime and fatigue limit stresses for different testing temperatures and environments, and compared the fatigue damage evolution between C/SiC and SiC/SiC composite through the fatigue hysteresis dissipated energy and fiber/matrix interface shear stress degradation rate. Li [23] developed the hysteresis loops model of fiber-reinforced CMCs considering oxidation subjected to static fatigue loading. Pailler and Lamon [24] investigated the static fatigue behavior of SiC/SiC minicomposite considering the interphase degradation at elevated temperatures. For the lifetime prediction of CMCs, besides the interphase degradation, the fiber strength degradation should also be considered. Under dwell-fatigue loading at elevated temperature in the oxidative environment, the fiber/matrix interface debonding and sliding between fibers and the matrix are affected by the combination effects of interface oxidation and interface wear [25].

In this chapter, the damage development and fatigue lifetime prediction of cross-ply SiC/MAS composite subjected to the dwell-fatigue loading at elevated temperatures in an oxidizing atmosphere is investigated. The fatigue hysteresis-based damage models under dwell fatigue at elevated temperature are developed considering the fiber/matrix interface debonding, sliding, wear and oxidation. The fiber failure probability is determined in the fiber/matrix interface debonding region and interface bonding region considering fiber/matrix interface wear and oxidation. Relationships between the damage parameters and internal damage development of matrix cracking, fiber/matrix interface debonding and sliding, and fiber fracture are established. Experimental damage development and fatigue lifetime curves of cross-ply Nicalon™ SiC/MAS subjected to dwell-fatigue loading at elevated temperature in air are predicted.

6.2 MICROMECHANICAL HYSTERESIS CONSTITUTIVE MODEL

Under dwell-fatigue loading, the damage development inside of the fiber-reinforced CMCs depends on the dwell-fatigue duration, cyclic number, and fatigue peak stress. In the present analysis, the combining effects of the fiber/matrix interface oxidation and interface wear in the interface debonding region on the interface debonding, interface sliding, and fiber fracture in different damage regions are considered. Relationships between the fatigue hysteresis-based damage parameters, dwell-fatigue duration, cycle number, and fatigue peak stress are established based on the damage development models. Based on the fiber/matrix interface debonding and interface sliding between the fiber and the matrix inside of composite, the fatigue hysteresis loops of fiber-reinforced CMCs subjected to dwell-fatigue loading can be divided into four different cases:

Case 1: the fiber/matrix interface oxidation region and the interface wear region are less than matrix crack spacing, and the fiber/matrix interface counter-slip and the fiber/matrix interface new-slip lengths are equal to the fiber/matrix interface debonded length.

Case 2: the fiber/matrix interface oxidation region and the fiber/matrix interface wear region are less than matrix crack spacing, and the fiber/matrix interface counter-slip and the interface new-slip lengths are less than the fiber/matrix interface debonded length.

Case 3: the fiber/matrix interface oxidation region and the interface wear region are equal to matrix crack spacing, and the fiber/matrix interface counter-slip and the interface new-slip lengths are less than matrix crack spacing.

Case 4: the fiber/matrix interface oxidation region and the interface wear region are equal to matrix crack spacing, and the fiber/matrix interface counter-slip and the interface new-slip lengths are equal to matrix crack spacing.

When the fiber/matrix interface oxidation region and the interface wear region are less than matrix crack spacing, the unloading and reloading stress-strain relationships are determined using the following equations:

$$
\varepsilon_{c\_pu} = \frac{2\sigma L_d}{V_f E_f L_c} + \frac{2\tau_f}{r_f E_f L_c}\xi^2 + \frac{4\tau_f}{r_f E_f L_c}\xi\left(L_d - \xi\right)
$$

$$
+ \frac{4\tau_i(N)}{r_f E_f L_c}\left(L_y - \xi\right)^2 - \frac{2\tau_i(N)}{r_f E_f L_c}\left(2L_y - \xi - L_d\right)^2 + \frac{2\sigma_{fo}}{E_f L_c}\left(\frac{L_c}{2} - L_d\right)
$$

$$
+ \frac{2r_f}{\rho E_f L_c}\left[\frac{V_m}{V_f}\sigma_{mo} + \frac{2\tau_f}{r_f}\xi + \frac{2\tau_i(N)}{r_f}\left(2L_y - \xi - L_d\right)\right]
$$

$$
\times\left[1 - \exp\left(-\rho\frac{L_c/2 - L_d}{r_f}\right)\right] - \left(\alpha_c - \alpha_f\right)\Delta T. \tag{6.1}
$$

$$
\varepsilon_{c\_pr} = \frac{2\sigma}{V_f E_f L_c}L_d - \frac{4\tau_f}{r_f E_f L_c}L_z^2 + \frac{2\tau_f}{r_f E_f L_c}\left(2L_z - \xi\right)^2
$$

$$
- \frac{4\tau_f}{r_f E_f L_c}\left(2L_z - \xi\right)\left(L_d - \xi\right) + \frac{4\tau_i(N)}{r_f E_f L_c}\left(L_y - \xi\right)^2
$$

$$
- \frac{2\tau_i(N)}{r_f E_f L_c}\left(2L_y - \xi - L_d\right)^2 + \frac{2\sigma_{fo}}{E_f L_c}\left(\frac{L_c}{2} - L_d\right)
$$

$$
+ \frac{2r_f}{\rho E_f L_c}\left[\frac{V_m}{V_f}\sigma_{mo} - \frac{2\tau_f}{r_f}\left(2L_z - \xi\right) + \frac{2\tau_i(N)}{r_f}\left(2L_y - \xi - L_d\right)\right] \tag{6.2}
$$

$$
\times\left[1 - \exp\left(-\rho\frac{L_c/2 - L_d}{r_f}\right)\right] - \left(\alpha_c - \alpha_f\right)\Delta T,
$$

where σ denotes the applied stress; V_f and V_m denote the fiber and matrix volume content, respectively; E_f denotes the fiber elastic modulus; ξ is the interface oxidation length; L_c denotes the matrix crack spacing; τ_f denotes the fiber/matrix interface shear stress in the interface oxidation region; σ_{fo} and σ_{mo} denote the fiber and matrix axial stress in the fiber/matrix interface bonded region, respectively; ρ denotes the shear-lag model parameter; α_f and α_c denote the fiber and composite thermal expansion coefficient, respectively; and ΔT denotes the temperature difference between the fabricated temperature T_0 and testing temperature T_1 ($\Delta T = T_1 - T_0$). The interface counter-slip length L_y and interface new slip length L_z considering the combining effects of the fiber/matrix interface oxidation and interface wear are determined using the following equations when the fiber/matrix interface partially debonds.

$$
L_y = \frac{1}{2}\left\{ L_d + \left(1 - \frac{\tau_f}{\tau_i(N)}\right)\xi \right.
$$

$$
\left. - \left[\frac{r_f}{2}\left(\frac{V_m E_m}{V_f E_c}\frac{\sigma}{\tau_i(N)} - \frac{1}{\rho}\right) - \sqrt{\left(\frac{r_f}{2\rho}\right)^2 + \frac{r_f V_m E_m E_f}{E_c\left[\tau_i(N)\right]^2}\Gamma_i} \right] \right\} \tag{6.3}
$$

$$
L_z = \frac{\tau_i(N)}{\tau_f}\left\{ L_y - \frac{1}{2}\left[L_d + \left(1 - \frac{\tau_f}{\tau_i(N)}\right)\xi \right.\right.
$$

$$
\left.\left. - \left[\frac{r_f}{2}\left(\frac{V_m E_m}{V_f E_c}\frac{\sigma}{\tau_i(N)} - \frac{1}{\rho}\right) - \sqrt{\left(\frac{r_f}{2\rho}\right)^2 + \frac{r_f V_m E_m E_f}{E_c\left[\tau_i(N)\right]^2}\Gamma_i} \right] \right]\right\}, \tag{6.4}
$$

where Γ_i is the interface debonding energy.

When the fiber/matrix interface oxidation region and the interface wear region are equal to matrix crack spacing, the unloading and reloading stress-strain relationships are determined using the following equations:

$$\varepsilon_{c\_fu} = \frac{\sigma}{V_f E_f} - \frac{2\tau_f}{r_f E_f L_c}\xi^2 + \frac{2\tau_f}{r_f E_f}\xi + \frac{4\tau_i(N)}{r_f E_f L_c}(L_y - \xi)^2$$

$$-\frac{2\tau_i(N)}{r_f E_f L_c}\left(2L_y - \xi - \frac{L_c}{2}\right)^2 - (\alpha_c - \alpha_f)\Delta T, \tag{6.5}$$

$$\varepsilon_{c\_fr} = \frac{\sigma}{V_f E_f} - \frac{4\tau_f}{r_f E_f L_c}L_z^2 + \frac{2\tau_f}{r_f E_f L_c}(2L_z - \xi)^2$$

$$-\frac{4\tau_f}{r_f E_f L_c}(2L_z - \xi)(L_y - \xi) + \frac{4\tau_i(N)}{r_f E_f L_c}(L_y - \xi)^2$$

$$-\frac{4\tau_f}{r_f E_f L_c}(2L_z - \xi)\left(\frac{L_c}{2} - L_y\right) \tag{6.6}$$

$$-\frac{2\tau_i(N)}{r_f E_f L_c}\left(2L_y - \xi - \frac{L_c}{2}\right)^2 - (\alpha_c - \alpha_f)\Delta T,$$

where the fiber/matrix interface counter-slip length and interface new-slip length considering the combining effects of the interface oxidation and interface wear are determined using the following equation, corresponding to the fiber/matrix interface completely debonding:

$$L_y = \left[1 - \frac{\tau_f}{\tau_i(N)}\right]\xi + \frac{r_f V_m E_m}{4V_f E_c \tau_i(N)}(\sigma_{max} - \sigma), \tag{6.7}$$

$$L_z = L_y(\sigma_{min}) - \frac{r_f V_m E_m}{4V_f E_c \tau_i(N)}(\sigma_{max} - \sigma), \tag{6.8}$$

where σ_{max} and σ_{min} denote the peak and valley stress, respectively.

The fatigue hysteresis dissipated energy ΔW is defined using the following equation:

$$\Delta W = \int_{\sigma_{min}}^{\sigma_{max}} \left[\varepsilon_{c\_unload}(\sigma) - \varepsilon_{c\_reload}(\sigma)\right]d\sigma, \tag{6.9}$$

where ε_{c_unload} and ε_{c_reload} denote the unloading and reloading strain, respectively.

6.3 MICROMECHANICAL LIFETIME PREDICTION MODEL

Under dwell-fatigue loading, fiber failure in fiber-reinforced CMCs depends on the degradation of fiber strength in different damage regions, that is, the fiber/matrix interface oxidation region, interface wear region, and interface debonding region. In the present analysis, the fiber failure probabilities in different damage regions are determined considering the dwell-fatigue duration, cyclic number, and fatigue peak stress. The fibers broken fraction versus cycle number curves and fatigue lifetime subjected to dwell-fatigue loading are predicted. The global load sharing (GLS) assumption is used to determine the load carried by intact and fracture fibers.

$$\frac{\sigma}{V_f} = \left[1 - P_f\left(1 + \frac{2L_f}{L_c}\right)\right]\Phi + P_r\frac{2L_f}{L_c}\langle\Phi_b\rangle, \qquad (6.10)$$

where L_f denotes the slip length over which the fiber stress would decay to zero if not interrupted by the far-field equilibrium stresses, $\langle\Phi_b\rangle$ denotes the average stress carried by broken fibers, P_f denotes the total fiber failure probability, and P_r denotes the fiber failure probability in the interface debonding region and interface bonded region.

When matrix cracking and interface debonding occur subjected to dwell-fatigue loading, the micromechanical stress distribution in fibers changes along the fiber length in different damage regions. Due to the characteristic of statistical failure for fibers, the total fiber failure probability P_f is determined as a sum of fiber failure probability in the interface oxidation region, interface debonding region, and interface bonding region, following the treatment of fiber fracture probability in different damage regions by Oh and Finnie [26], Curtin [27], Liao and Reifsnider [28], and Li [29].

The total fiber failure probability P_f and the fiber failure probability in the interface debonding region and interface bonded region P_r are determined using the following equations:

$$P_f = \chi\left[\zeta P_{fa} + (1 - \eta)P_{fb}\right] + P_{fc} + P_{fd}, \qquad (6.11)$$

$$P_r = P_{fc} + P_{fd}, \tag{6.12}$$

where P_{fa}, P_{fb}, P_{fc}, and P_{fd} denote the fiber failure probability of oxidized fibers in the oxidation region, unoxidized fibers in the oxidation region, fibers in the interface debonded region, and interface bonded region, respectively; ζ denotes the oxidation fiber fraction in the oxidized region; and χ denotes the fraction of oxidation in the multiple matrix cracks.

$$P_{fa}(\Phi) = 1 - \exp\left\{ -2\frac{L_t}{l_0}\left[\frac{\Phi}{\sigma_0(t)}\right]^{m_f} \right\}, \tag{6.13}$$

$$P_{fb}(\Phi) = 1 - \exp\left\{ -2\frac{L_t}{l_0}\left(\frac{\Phi}{\sigma_0}\right)^{m_f} \right\}, \tag{6.14}$$

$$P_{fc}(\Phi) = 1 - \exp\left\{ -\frac{r_f\Phi^{m_f+1}}{l_0(\sigma_0(N))^{m_f}\tau_i(N)(m_f+1)}\left[1 - \left(1 - \frac{L_d(N)}{L_f(N)}\right)^{m_f+1}\right] \right\}, \tag{6.15}$$

$$P_{fd}(\Phi) = 1 - \exp\left\{ -\frac{2r_f\Phi^{m_f}}{\rho l_0(\sigma_0(N))^{m_f}(m_f+1)\left(1 - \frac{\sigma_{fo}}{\Phi} - \frac{L_d(N)}{L_f(N)}\right)} \right.$$
$$\times \left[\left(1 - \frac{L_d(N)}{L_f(N)} - \left(1 - \frac{\sigma_{fo}}{\Phi} - \frac{L_d(N)}{L_f(N)}\right)\frac{\rho L_d(N)}{r_f}\right)^{m_f+1}\right.$$
$$\left.\left. - \left(1 - \frac{L_d(N)}{L_f(N)} - \left(1 - \frac{\sigma_{fo}}{\Phi} - \frac{L_d(N)}{L_f(N)}\right)\frac{\rho L_c}{2r_f}\right)^{m_f+1}\right] \right\}, \tag{6.16}$$

where Φ denotes the load carried by intact fibers and σ_{fo} denotes the fiber stress in the interface bonded region. The time-dependent fiber strength is controlled by surface defects resulting from oxidation [30].

$$\sigma_0(t) = \begin{cases} \sigma_0, t \le \dfrac{1}{k}\left(\dfrac{K_{IC}}{Y\sigma_0}\right)^4 \\ \dfrac{K_{IC}}{Y\sqrt[4]{kt}}, t > \dfrac{1}{k}\left(\dfrac{K_{IC}}{Y\sigma_0}\right)^4 \end{cases}, \tag{6.17}$$

where K_{IC} denotes the critical stress intensity factor, Y is a geometric parameter, and k is the parabolic rate constant.

With increasing of applied cycle number, the fiber/matrix interface shear stress and fibers strength decrease due to the fiber/matrix interface wear and interface oxidation. The fiber failure probability in the fiber/matrix interface oxidation region, interface debonding region, and interface bonded region can be obtained by combining the interface wear model, interface oxidation model, and fiber strength degradation model. The evolution of fiber failure probability versus applied cycle number curves can be obtained. When the fibers broken fraction approaches the critical value, the composite fatigue fractures.

6.4 EXPERIMENTAL COMPARISONS

Grant [31] investigated the cyclic tension-tension fatigue behavior of cross-ply Nicalon™ SiC/MAS composite under dwell-fatigue loading at $T = 566$ and 1093 °C in air atmosphere. The fatigue tests were conducted under the load control. The loading frequency was $f = 1$ Hz, and the fatigue load ratio (i.e., minimum to maximum stress) was $R = 0.1$. At $T = 566°C$ in air condition, the composite tensile strength was approximately $\sigma_{UTS} = 292$ MPa, and the fatigue peak stresses were $\sigma_{max} = 137$ and 103 MPa, with the dwell-fatigue duration $t = 1, 10,$ and 100 s, and at $T = 1093°C$ in air condition, the composite tensile strength were approximately $\sigma_{uts} = 209$ MPa, and the fatigue peak stresses were $\sigma_{max} = 137$ MPa and 103 MPa, with the dwell-fatigue duration $t = 1, 10,$ and 100 s.

6.4.1 Cross-Ply SiC/MAS at 566°C in an Air Condition

Under dwell-fatigue peak stress $\sigma_{max} = 137$ MPa with the dwell-fatigue duration $t = 1, 10,$ and 100 s, the experimental fatigue hysteresis dissipated energy (ΔW) and the normalized fatigue hysteresis modulus (E_{NOR}) decrease with applied cycles, and the fatigue peak strain (ε_p) increases with applied cycles, as shown in Figure 6.1 and Table 6.1.

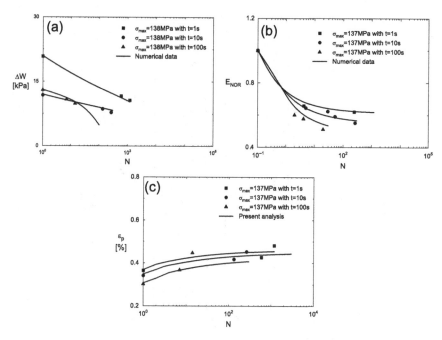

FIGURE 6.1 (a) Experimental and predicted fatigue hysteresis dissipated energy, (b) normalized experimental and predicted fatigue hysteresis modulus, and (c) experimental and predicted fatigue peak strain of cross-ply Nicalon™ SiC/ MAS composite under σ_{max} = 137 MPa at T = 566°C with the dwell-fatigue duration t = 1, 10, and 100 s.

When the dwell-fatigue duration is t = 1 s, the experimental fatigue hysteresis dissipated energy decreases from ΔW = 20.8 kPa at the first applied cycle to ΔW = 10.6 kPa at the 1195th applied cycle due to matrix cracking, interface debonding and sliding, and the theoretical predicted fatigue hysteresis dissipated energy decreases from ΔW = 20.9 kPa at the first applied cycle to ΔW = 10.2 kPa at the 1195th applied cycle, corresponding to the fiber/matrix interface slip Case 4; that is, the fiber/matrix interface oxidation region and the interface wear region are equal to the matrix crack spacing, and the fiber/matrix interface counter-slip and the interface new-slip lengths are equal to the matrix crack spacing. The normalized experimental fatigue hysteresis modulus (E_{NOR}) decreases rapidly at the initial applied cycles due to matrix cracking and fiber/matrix interface debonding, that is, from E_{NOR} = 1.0 on first loading to E_{NOR} = 0.61 at the 247th applied cycle, and the theoretical predicted normalized fatigue hysteresis modulus (E_{NOR}) decreases from E_{NOR} = 1.0 upon first loading to

TABLE 6.1 Damage Development of Cross-ply Nicalon™ SiC/MAS Composite Subjected to Dwell-Fatigue Loading at $\sigma_{max} = 137$ MPa and $T = 566°C$ in Air Condition

$t = 1$ s

Items	Experiment		Theory	
ΔW/(kPa)	$N=1$	$N=1195$	$N=1$	$N=1195$
	20.8	10.6	20.9	10.2

Items	Experiment		Theory	
E_{NOR}	$N=1$	$N=247$	$N=1$	$N=1195$
	1.0	0.61	1.0	0.61

Items	Experiment		Theory	
ε_p/(%)	$N=1$	$N=1195$	$N=1$	$N=1195$
	0.365	0.482	0.369	0.455

$t = 10$ s

Items	Experiment		Theory	
ΔW/(kPa)	$N=1$	$N=265$	$N=1$	$N=3000$
	11.8	7.8	11.9	8.2

Items	Experiment		Theory	
E_{NOR}	$N=1$	$N=263$	$N=1$	$N=300$
	1.0	0.55	1.0	0.56

Items	Experiment		Theory	
ε_p/(%)	$N=1$	$N=265$	$N=1$	$N=3000$
	0.342	0.453	0.349	0.443

$t = 100$ s

Items	Experiment		Theory	
ΔW/(kPa)	$N=1$	$N=14$	$N=1$	$N=100$
	13	9.8	13	4.8

Items	Experiment		Theory	
E_{NOR}	$N=1$	$N=20$	$N=1$	$N=30$
	1.0	0.51	1.0	0.53

Items	Experiment		Theory	
ε_p/(%)	$N=1$	$N=14$	$N=1$	$N=300$
	0.303	0.448	0.307	0.407

$E_{NOR} = 0.61$ at the 1195th applied cycle. The experimental fatigue peak strain (ε_p) increases from $\varepsilon_p = 0.365\%$ at the first applied cycle to $\varepsilon_p = 0.482\%$ at the 1195th applied cycle due to matrix cracking and fiber/matrix interface debonding and sliding, and the theoretical predicted fatigue peak strain (ε_p) increases from $\varepsilon_p = 0.369\%$ at the first applied cycle to $\varepsilon_p = 0.455\%$ at the 1195th applied cycle.

When the dwell-fatigue duration is $t = 10$ s, the experimental fatigue hysteresis dissipated energy decreases from $\Delta W = 11.8$ kPa at the first applied cycle to $\Delta W = 7.8$ kPa at the 265th applied cycle due to matrix cracking and fiber/matrix interface debonding and sliding, and the theoretical predicted fatigue hysteresis dissipated energy decreases from $\Delta W = 11.9$ kPa at the first applied cycle to $\Delta W = 8.2$ kPa at the 300th applied cycle, corresponding to the fiber/matrix interface slip Case 4; that is, the fiber/matrix interface completely debonding and the fiber completely sliding relative to the matrix in the interface debonding region. The normalized experimental fatigue hysteresis modulus (E_{NOR}) decreases from $E_{NOR} = 1.0$ upon first loading to $E_{NOR} = 0.55$ at the 263th applied cycle due to matrix cracking and fiber/matrix interface debonding, and the theoretical predicted fatigue hysteresis modulus (E_{NOR}) decreases from $E_{NOR} = 1.0$ upon first loading to $E_{NOR} = 0.56$ at the 300th applied cycle. The experimental fatigue peak strain (ε_p) increases from $\varepsilon_p = 0.342\%$ at the first applied cycle to $\varepsilon_p = 0.453\%$ at the 265th applied cycle due to matrix cracking and fiber/matrix interface debonding and sliding, and the theoretical predicted fatigue peak strain (ε_p) increases from $\varepsilon_p = 0.349\%$ at the first applied cycle to $\varepsilon_p = 0.443\%$ at the 3000th applied cycle.

When the dwell-fatigue duration is $t = 100$ s, the experimental fatigue hysteresis dissipated energy decreases from $\Delta W = 13$ kPa at the first applied cycle to $\Delta W = 9.8$ kPa at the 14th applied cycle due to matrix cracking and fiber/matrix interface debonding and sliding, and the theoretical predicted fatigue hysteresis dissipated energy decreases from $\Delta W = 13$ kPa at the first applied cycle to $\Delta W = 4.8$ kPa at the 100th applied cycle, corresponding to the fiber/matrix interface slip Case 4; that is, the fiber/matrix interface completely debonding and the fiber completely sliding relative to the matrix in the interface debonding region. The normalized experimental fatigue hysteresis modulus (E_{NOR}) decreases from $E_{NOR} = 1.0$ on first loading to $E_{NOR} = 0.51$ at the 20th applied cycle due to matrix cracking and fiber/matrix interface debonding, and the theoretical predicted fatigue hysteresis modulus (E_{NOR}) decreases from $E_{NOR} = 1.0$ on first loading to $E_{NOR} = 0.53$ at the 30th applied cycle. The experimental fatigue peak strain (ε_p) increases from $\varepsilon_p = 0.303\%$ at the first applied cycle to $\varepsilon_p = 0.448\%$ at the 14th applied cycle due to matrix cracking and fiber/matrix interface debonding and sliding, and the theoretical predicted fatigue peak strain (ε_p) increases from $\varepsilon_p = 0.307\%$ at the first applied cycle to $\varepsilon_p = 0.407\%$ at the 300th applied cycle.

The experimental and theoretical predicted fatigue lifetime curves and the theoretical predicted broken fibers versus applied cycles curves under $\sigma_{max} = 137$ MPa of cross-ply Nicalon™ SiC/MAS composite with the dwell-fatigue duration $t = 1$, 10, and 100 s at $T = 566$ °C in air condition are shown in Figure 6.2. With an increase in dwell-fatigue time, the fatigue lifetime under the high fatigue peak stress rapidly decreases, and the fatigue limit stress decreases. Experimental and predicted failure cycles of cross-ply Nicalon™ SiC/MAS composite under dwell-fatigue loading at $T = 566$ °C in an air condition are shown in Table 6.2.

When the dwell-fatigue duration is $t = 1$ s, the theoretical predicted broken fiber fraction increases from $P = 0.5\%$ at the first applied cycle to $P = 31.2\%$ at the 1917th applied cycle; when the dwell-fatigue duration is $t = 10$ s, the theoretical predicted broken fiber fraction increases from $P = 0.5\%$ at the first applied cycle to $P = 31.1\%$ at the 359th applied cycle; and when the dwell-fatigue duration is $t = 100$ s, the theoretical predicted broken fiber fraction increases from $P = 0.6\%$ at the first applied cycle to $P = 30.8\%$ at the 22nd applied cycle. In the present analysis, the GLS criterion was adopted to determine the load carrying between intact and broken fibers, leading to the high fiber failure probability under cyclic fatigue loading.

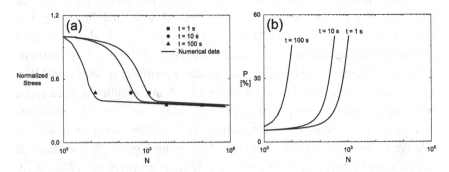

FIGURE 6.2 (a) Experimental and predicted fatigue life S–N curves and (b) the predicted broken fiber fraction versus applied cycles curves under $\sigma_{max} = 137$ MPa of cross-ply Nicalon™ SiC/MAS composite with the dwell-fatigue duration $t = 1$, 10, and 100 s at $T = 566$°C in air condition.

TABLE 6.2   Experimental and Predicted Failure Cycles of Cross-Ply Nicalon™ SiC/MAS Composite Subjected to Dwell-Fatigue Loading at 566°C in Air Condition

	$t = 1$ s				$t = 10$ s				$t = 100$ s			
	Experiment		Theory		Experiment		Theory		Experiment		Theory	
Items	σ_{max} = 137 MPa	σ_{max} = 103 MPa	σ_{max} = 137 MPa	σ_{max} = 103 MPa	σ_{max} = 137 MPa	σ_{max} = 103 MPa	σ_{max} = 137 MPa	σ_{max} = 103 MPa	σ_{max} = 137 MPa	σ_{max} = 103 MPa	σ_{max} = 137 MPa	σ_{max} = 103 MPa
Cycles	1195	101834	1934	1181930	265	91739	362	92868	14	5135	24	11983

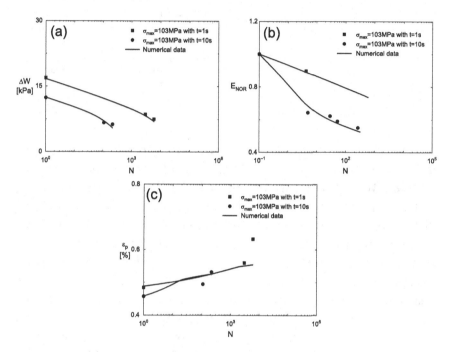

FIGURE 6.3   (a) Experimental and predicted fatigue hysteresis dissipated energy, (b) normalized experimental and predicted fatigue hysteresis modulus, and (c) experimental and predicted fatigue peak strain of cross-ply Nicalon™ SiC/MAS composite under σ_{max} = 103 MPa at T = 1093°C with dwell-fatigue duration $t = 1$ and 10 s.

6.4.2   Cross-Ply SiC/MAS at 1093°C in an Air Condition

Under dwell-fatigue peak stress σ_{max} = 103 MPa with the dwell-fatigue duration $t = 1$ and 10 s, the experimental fatigue hysteresis dissipated energy (ΔW) and normalized fatigue hysteresis modulus (E_{NOR}) decrease with applied cycles, and the fatigue peak strain (ε_p) increases with applied cycles, as shown in Figure 6.3 and Table 6.3.

TABLE 6.3 Damage Development of Cross-ply Nicalon™–SiC/MAS Composite Subjected to Dwell-Fatigue Loading at σ_{max} = 103 MPa and T = 1093°C in Air Condition

	$t = 1$ s				$t = 10$ s			
	Experiment		Theory		Experiment		Theory	
Items	$N = 1$	$N = 6017$	$N = 1$	$N = 6017$	$N = 1$	$N = 216$	$N = 1$	$N = 216$
ΔW/(kPa)	16.9	7.5	16.7	6.9	12.4	6.3	12.4	5.4

	$t = 1$ s				$t = 10$ s			
	Experiment		Theory		Experiment		Theory	
Items	$N = 1$	$N = 4$	$N = 1$	$N = 600$	$N = 1$	$N = 263$	$N = 1$	$N = 316$
E_{NOR}	1.0	0.89	1.0	0.73	1.0	0.55	1.0	0.52

	$t = 1$ s				$t = 10$ s			
	Experiment		Theory		Experiment		Theory	
Items	$N = 1$	$N = 6017$	$N = 1$	$N = 6017$	$N = 1$	$N = 216$	$N = 1$	$N = 216$
ε_p/(%)	0.484	0.633	0.487	0.55	0.457	0.532	0.457	0.525

When the dwell-fatigue duration is $t = 1$ s, the experimental fatigue hysteresis dissipated energy decreases from ΔW = 16.9 kPa at the first applied cycle to ΔW = 7.5 kPa at the 6017th applied cycle due to matrix cracking and fiber/matrix interface debonding and sliding, and the theoretical predicted fatigue hysteresis dissipated energy decreases from ΔW = 16.7 kPa at the first applied cycle to ΔW = 6.9 kPa at the 6017th applied cycle, corresponding to the fiber/matrix interface slip Case 4; that is, the fiber/matrix interface complete debonding and the fiber complete sliding relative to the matrix in the interface debonding region. The normalized experimental fatigue hysteresis modulus (E_{NOR}) decreases rapidly at the initial applied cycles due to matrix cracking and fiber/matrix interface debonding, that is, from E_{NOR} =1.0 on first loading to E_{NOR} = 0.89 at the 4th applied cycle, and the normalized theoretical predicted fatigue hysteresis modulus (E_{NOR}) decreases from E_{NOR} =1.0 on first loading to E_{NOR} = 0.73 at the 600th applied cycle. Experimental fatigue peak strain (ε_p) increases from ε_p = 0.484% at the first applied cycle to ε_p = 0.633% at the 6017th applied cycle due to matrix cracking and fiber/matrix interface debonding and sliding, and the theoretical predicted fatigue peak strain (ε_p) increases from ε_p = 0.487% at the first applied cycle to ε_p = 0.55% at the 6017th applied cycle.

When the dwell-fatigue duration is $t = 10$ s, the experimental fatigue hysteresis dissipated energy decreases from ΔW = 12.4 kPa at the first applied cycle to ΔW = 6.3 kPa at the 216th applied cycle due to matrix cracking and fiber/matrix interface debonding and sliding, and the theoretical predicted fatigue hysteresis dissipated energy decreases from $\Delta W =$ 12.4 kPa at the first applied cycle to $\Delta W =$ 5.4 kPa at the 216th applied cycle, corresponding to the fiber/matrix interface slip Case 4; that is, the fiber/matrix interface completely debonding and the fiber completely sliding relative to the matrix in the interface debonding region. The normalized experimental fatigue hysteresis modulus (E_{NOR}) decreases from $E_{NOR} = 1.0$ on first loading to $E_{NOR} = 0.55$ at the 263th applied cycle due to matrix cracking and fiber/matrix interface debonding, and the theoretical predicted fatigue hysteresis modulus (E_{NOR}) decreases from $E_{NOR} = 1.0$ upon first loading to $E_{NOR} = 0.52$ at the 316th applied cycle. The experimental fatigue peak strain (ε_p) increases from $\varepsilon_p = 0.457\%$ at the first applied cycle to $\varepsilon_p = 0.532\%$ at the 216th applied cycle due to matrix cracking and fiber/matrix interface debonding and sliding, and the theoretical predicted fatigue peak strain (ε_p) increases from $\varepsilon_p = 0.457\%$ at the first applied cycle to $\varepsilon_p = 0.525\%$ at the 216th applied cycle.

Experimental and predicted fatigue lifetime curves and the theoretical predicted broken fiber fraction versus applied cycle curves of cross-ply Nicalon™ SiC/MAS composite under $\sigma_{max} = 103$ MPa, with the dwell-fatigue duration $t = 1$, 10, and 100 s at 1093°C in air condition, are shown in Figure 6.4. With increasing dwell-fatigue duration, the fatigue lifetime under high fatigue peak stress rapidly decreases, and the fatigue limit stress decreases. When the dwell-fatigue duration is $t = 1$ s, the fatigue limit stress approaches 28% tensile strength; when the dwell-fatigue duration is $t = 10$ s, the fatigue limit stress approaches 10% tensile strength; and when the dwell-fatigue duration is $t = 100$ s, the fatigue limit stress approaches 8% tensile strength. Experimental and predicted failure cycles of cross-ply Nicalon™–SiC/MAS composite subjected to dwell-fatigue loading at 1093°C in air condition are shown in Table 6.4.

Under $\sigma_{max} = 103$ MPa, the theoretical predicted broken fiber fraction increases from $P = 0.7\%$ at the first applied cycle to $P = 31.2\%$ at the 3076th applied cycle when the dwell-fatigue duration is $t = 1$ s; when the dwell-fatigue duration is $t = 10$ s, the theoretical predicted broken fiber fraction increases from $P = 0.7\%$ at the first applied cycle to $P = 31.2\%$ at the 259th applied cycle; and when the dwell-fatigue duration is $t = 100$ s, the theoretical predicted broken fiber fraction increases from $P = 0.5\%$ at the first applied cycle to $P = 31.1\%$ at the 22nd applied cycle.

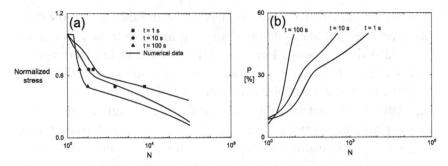

FIGURE 6.4 (a) Experimental and predicted fatigue life S–N curves and (b) predicted broken fiber fraction versus applied cycles curves under σ_{max} = 103 MPa of cross-ply Nicalon™ SiC/MAS composite with the dwell-fatigue duration t = 1, 10, and 100 s at T = 1093°C in air condition.

TABLE 6.4 Experimental and Predicted Failure Cycles of Cross-Ply Nicalon™ SiC/MAS Composite Subjected to Dwell-Fatigue Loading at T = 1093°C in Air Condition

	t = 1 s				t = 10 s				t = 100 s			
	Experiment		Theory		Experiment		Theory		Experiment		Theory	
Items	σ_{max} = 137 MPa	σ_{max} = 103 MPa	σ_{max} = 137 MPa	σ_{max} = 103 MPa	σ_{max} = 137 MPa	σ_{max} = 103 MPa	σ_{max} = 137 MPa	σ_{max} = 103 MPa	σ_{max} = 137 MPa	σ_{max} = 103 MPa	σ_{max} = 137 MPa	σ_{max} = 103 MPa
Cycles	18	6017	29	3237	11	216	13	240	4	10	5	24

6.5 SUMMARY AND CONCLUSION

In this chapter, the damage development and fatigue lifetime prediction of cross-ply SiC/MAS subjected to dwell-fatigue loading at elevated temperatures in an oxidizing atmosphere were investigated. Relationships between the fatigue hysteresis-based damage parameters and internal composite damage development of matrix cracking, fiber/matrix interface debonding and sliding, and fiber fracture were established. Experimental fatigue life curves of cross-ply Nicalon™–SiC/MAS under dwell-fatigue loading at elevated temperatures in an air condition were predicted. With an increase in dwell-fatigue time, the degradation rate of fatigue hysteresis dissipated energy and fatigue hysteresis modulus, and the increasing rate of fatigue peak strain all increase; however, the fatigue lifetime decreases. The comparisons between the theoretical models proposed in the present analysis and experimental results proved the validity of the damage evolution (i.e., hysteresis dissipated energy, hysteresis modulus, and peak strain) and the fatigue lifetime prediction.

REFERENCES

1. Naslain R. Design, preparation and properties of non-oxide CMCs for application in engines and nuclear reactors: an overview. *Compos. Sci. Technol.* 2004; 64:155–170.
2. Santhosh U, Ahmad J, Ojard G, Miller R, Gowayed Y. Deformation and damage modeling of ceramic matrix composites under multiaxial stresses. *Compos. Part B* 2016; 90:97–106.
3. Triantou KI, Mergia K, Perez B, Florez S, Stefan A, Ban C, Pelin G, Ionescu G, Zuber C, Fischer WPP, Barcena J. Thermal shock performance of carbon-bonded fiber composite and ceramic matrix composite joints for thermal protection re-entry applications. *Compos. Part B* 2017; 111:270–278.
4. CCAR 33-R2, Airworthiness regulations of aeroengine. Civil Aviation Administration of China, 2011.
5. FAA AC 21-23B, Airworthiness certification of civil aircraft, engine, propellers and related products imported to the United States. U.S. Department of Transportation, Federal Aviation Administration, 2004.
6. Santhosh U, Ahmad J, John R, Ojard G, Miller R, Gowayed Y. Modeling of stress concentration in ceramic matrix composites. *Compos. Part B* 2013; 45:1156–1163.
7. Li LB. Modeling hysteresis behavior of cross-ply C/SiC ceramic matrix composites. *Compos. Part B* 2013; 53:36–45.
8. Zhu SJ, Mizuno M, Nagano Y, Cao JW, Kagawa Y, Kaya H. Creep and fatigue behavior in an enhanced SiC/SiC composite at high temperature. *J. Am. Ceram. Soc.* 1998; 81:2269–2277.
9. John R, Ojard G, Miler R, Gowayed Y, Morscher G, Santhosh U, Ahmad J. Frequency and hold-time effects on durability of melt-infiltrated SiC/SiC. *Ceram. Eng. Sci. Proc.* 2011; 32:101–109.
10. Gowayed Y, Ojard G, Chen J, Morscher G, Miller R, Santhosh U, Ahmad J, John R. Accumulation of time-dependent strain during dwell-fatigue experiments of Ibn-Sylramic melt infiltration SiC/SiC composites with and without holes. *Compos. Part A* 2011; 42:2020–2027.
11. Ruggles-Wrenn MB, Sharma V. Effects of steam environment on fatigue behavior of two SiC/[SiC+Si_3N_4] ceramic composite at 1300°C. *Appl. Compos. Mater.* 2011; 18:385–396.
12. Ruggles-Wrenn MB, Delapasse J, Chamberlain AL, Lane JE, Cook TS. Fatigue behavior of a Hi-Nicalon™/SiC–B_4C composite at 1200°C in air and in steam. *Mater. Sci. Eng. A* 2012; 534:119–128.
13. Ruggles-Wrenn MB, Lee MD. Fatigue behavior of an advanced SiC/SiC ceramic composite with a self-healing matrix at 1300°C in air and in steam. *Mater. Sci. Eng. A* 2016; 677:438–445.

14. Shi DQ, Jing X, Yang XG. Low cycle fatigue behavior of a 3D braided KD-I fiber reinforced ceramic matrix composite for coated and uncoated specimens at 1100°C and 1300°C. *Mater. Sci. Eng. A* 2015; 631:38–44.

15. Maillet E, Godin N, R'Mili M, Reynaud P, Fantozzi G, Lamon J. Real-time evaluation of energy attenuation: A novel approach to acoustic emission analysis for damage monitoring of ceramic matrix composites. *J. Eur. Ceram. Soc.* 2014; 34:1673–1679.

16. Singh YP, Mansour R, Morscher GN. Combined acoustic emission and multiple lead drop measurements in detailed examination of crack initiation and growth during interlaminar testing of ceramic matrix composites. *Compos. Part A* 2017; 97:93–99.

17. Zhang DY, Meyer P, Wass AM. An experimentally valiadated computational model for progressive damage analysis of notched oxide/oxide woven ceramic matrix composites. *Compos. Struct.* 2017; 161:264–274.

18. Li LB. A hysteresis dissipated energy-based parameter for damage monitoring of carbon fiber-reinforced ceramic-matrix composites under fatigue loading. *Mater. Sci. Eng. A* 2015; 634:188–201.

19. Li LB. A hysteresis dissipated energy-based damage parameter for life prediction of carbon fiber-reinforced ceramic-matrix composites under fatigue loading. *Compos. Part B* 2015; 82:108–128.

20. Li LB. Damage monitoring of unidirectional C/SiC ceramic-matrix composite under cyclic fatigue loading using a hysteresis loss energy-based damage parameter at room and elevated temperatures. *Appl. Compos. Mater.* 2016; 23:357–374.

21. Li LB. Fatigue damage and lifetime of SiC/SiC ceramic-matrix composite under cyclic loading at elevated temperatures. *Materials* 2017; 10:371.

22. Li LB. Comparisons of damage evolution between 2D C/SiC and SiC/SiC ceramic-matrix composites under tension-tension cyclic fatigue loading at room and elevated temperatures. *Materials* 2016; 9:844.

23. Li LB. Modeling the effect of oxidation on hysteresis loops of carbon fiber-reinforced ceramic-matrix composites under static fatigue at elevated temperatures. *J. Eur. Ceram. Soc.* 2016; 36:465–480.

24. Pailler F, Lamon J. Micromechanics based model of fatigue/oxidation for ceramic matrix composites. *Compos. Sci. Technol.* 2005; 65:369–374.

25. Li LB. Damage development and lifetime prediction of fiber-reinforced ceramic-matrix composites subjected to dwell-fatigue loading at elevated temperatures in oxidizing atmosphere. *J. Ceram. Soc. Japan* 2018; 126:516–528.

26. Oh HL, Finnie I. On the location of fracture in brittle solids I: due to static loading. *Int. J. Frac. Mech.* 1970; 6:287–300.

27. Curtin WA. Theory of mechanical properties of ceramic-matrix composites. *J. Am. Ceram. Soc.* 1991; 74:2837–2845.

28. Liao K, Reifsnider K. A tensile strength model for unidirectional fiber-reinforced brittle matrix composite. *Int. J. Frac.* 2000; 106:95–115.

29. Li LB. Modeling the effect of interface wear on fatigue hysteresis behavior of carbon fiber-reinforced ceramic-matrix composites. *Appl. Compos. Mater.* 2015; 22:887–920.

30. Lara-Curzio E. Analysis of oxidation-assisted stress-rupture of continuous fiber-reinforced ceramic matrix composites at intermediate temperatures. *Compos. Part A* 1999; 30:549–554.

31. Grant SA. Fatigue behavior of a cross-ply ceramic matrix composite subjected to tension-tension cycling with hold time. Master thesis, Air Force Institute of Technology, 1994.

CHAPTER 7

Mechanical Hysteresis Behavior in a Three-Dimensional Needle-Punched C/SiC Composite at Room Temperature

7.1 INTRODUCTION

Carbon fiber–reinforced composites (e.g., C/C, C/SiC, etc.) show excellent comprehensive properties at room and high temperatures, with high specific strength, specific modulus, thermal shock resistance, and oxidation resistance. As a reinforcement of composites, a carbon fiber preform has a decisive influence on the properties of composites. The three-dimensional (3D) preform contains load-bearing fibers in many directions, which overcomes the disadvantages of low damage tolerance and weak interlaminar performance of two-dimensional preform and shows higher load-bearing performance [1]. Traditional 3D carbon fiber–forming technology, such as braiding, weaving, and knitting, has a complex process and high cost, so it is not easy to mass-produce. The needling technology proposed by the French SPS company (SNECMA propulsion solid) can efficiently produce

DOI: 10.1201/b23026-7

various shapes of needled carbon fiber preforms. The needling technology is used to conduct relay needling on carbon fiber cloth, fiber web, and other fiber composite materials by needling, which leads part of the in-plane fibers into the ply thickness direction to produce vertical fiber clusters so that the carbon fiber cloth and fiber web are closely combined with each other to form a preform with certain strength in the plane and between layers. The cost of C/SiC composites based on needled carbon fiber preform is low, and it overcomes the disadvantage of weak interlaminar properties of 2D materials [2].

Many researchers performed theoretical and experimental investigations on the mechanical behavior of 3D needle-punched C/SiC composites. Nie et al. [3] performed experimental investigations on the tensile behavior of 3D needle-punched C/SiC composite. The tensile stress-strain curves exhibited obvious nonlinear behavior and can be divided into three regions: the initial linear region, nonlinear region, and quasi-linear region. The damage mechanisms of matrix cracking propagation, interface debonding, and fiber bridging and fracture contribute to the nonlinear behavior of 3D needle-punched C/SiC composites. Fang et al. [4] investigated the tensile damage accumulation of 3D needle-punched C/SiC composite at room temperature using acoustic emission. The initiation of tensile damage at room temperature occurs in the matrix due to the appearance of microcracks at the stress level of 40% of a composite's tensile strength. The existence of the pyrolytic carbon (PyC) interphase between the fiber and the matrix, and the pores in the matrix contributed to the toughness of the composite under tensile loading. Fan et al. [5] fabricated the 3D needle-punched C/SiC composite using the chemical vapor infiltration (CVI) and liquid melt infiltration method. The 3D needle-punched C/SiC composites were composed of the layers of 0° nonwoven fiber cloth, short fiber web, 90° nonwoven fiber cloth, and needled fibers with a great many pores and cracks. The composite's tensile strength along and perpendicular to the fiber's direction were $\sigma_{UTS} = 260$ and 118 MPa, respectively. Chen et al. [6] investigated high temperature tensile mechanical properties of 3D needle-punched C/SiC composites. The composite's tensile strength increased gradually from $\sigma_{UTS} = 98.7$ MPa at room temperature to $\sigma_{UTS} = 162.6$ MPa at $T = 1800$ °C and decreased to $\sigma_{UTS} = 154.3$ MPa at $T = 2000$ °C. At elevated temperatures, a large number of fibers were pulled out from the matrix, indicating that the interfacial shear stress decreased with increasing temperature. Nie et al. [7] performed an experimental study on the cyclic loading/unloading tensile behavior of 3D needle-punched C/SiC composite at room temperature. The composite's tensile strength was σ_{UTS} = 129.6 MPa, and the composite's fracture strain was ε_f = 0.61%. Under tensile loading, when the unloading peak stress was lower than σ_{max} = 80 MPa, the composite's residual strain and unloading modulus increased linearly with unloading peak stress. However, when the unloading peak stress was higher than σ_{max} = 80 MPa, the composite's residual strain and unloading modulus increased nonlinearly with unloading peak stress. Liu et al. [8] performed an experimental study on the monotonic and cyclic loading/unloading tensile behavior of different 3D needle-punched C/SiC composites. The relationship between microstructure damages and tensile mechanical behavior has been established. Mei and Cheng [9] compared the mechanical hysteresis behavior of different C/SiC composites (i.e., 3D needle-punched, 2D plain-woven, 2.5D woven, and 3D braided). It was found that the fibers preform affected the composite's ultimate tensile strength and hysteresis loops. Li [10] performed experimental investigations on the cyclic loading/unloading tensile behavior of unidirectional C/SiC composite at room and elevated temperatures. Damages in the matrix, interface, and fibers contributed to the evolution of the hysteresis loops with increasing peak stress. Guo et al. [11] performed the experimental investigations on the cyclic loading/unloading tensile damage evolution in 2D woven SiC/SiC composite and established the relationship between a composite's internal damage and natural frequency. Xie et al. [12] performed a theoretical investigation on the nonlinear constitutive relationship of 3D needle-punched C/SiC composite under shear loading. Residual strain existed on unloading, and the composite's unloading modulus decreased with increasing peak stress, mainly attributed to the damages of matrix cracking and fiber/matrix interface debonding. Li et al. [13, 14] developed a micromechanical constitutive relationship for unidirectional, 2D cross-ply, 2.5D woven, and 3D braided CMCs at room temperature. Li [15] analyzed the effect of stochastic loading on tensile damage and fracture of unidirectional, 2D cross-ply, and plain-woven CMCs. Li [16] developed a micromechanical tension-tension fatigue hysteresis loops models for unidirectional CMCs considering the fragmentation of the matrix.

In this chapter, the 3D needle-punched C/SiC composite was fabricated using the chemical vapor deposition (CVD) and reactive infiltration method. A micromechanical hysteresis constitutive relationship was developed considering damage mechanisms of matrix cracking, interface debonding and slip, and fiber fracture and pullout. Using the developed hysteresis loops models and damage models, the hysteresis loops of four different types of 3D needle-punched C/SiC composites were predicted for different peak stresses. Hysteresis parameters of unloading residual strain, peak strain, hysteresis loops width, hysteresis loops area, interface slip ratio, and inverse tangent modulus were adopted to characterize the tensile damage evolution inside of composites.

7.2 MATERIALS AND EXPERIMENTAL PROCEDURES

The reinforcement of needle-punched composite material is carbon cloth/short-chopped-fiber web needle-punched preform. The preform is composed of carbon fiber nonwoven cloth and short-chopped-fiber web layer. The nonwoven cloth is composed of unidirectional continuous long fiber bundles, and the fiber web is composed of short carbon fibers randomly distributed in different directions. Automatic production has already been realized in the forming process of needle punched preform:

- The 0° nonwoven cloth, fiber web, and 90° nonwoven cloth are alternately stacked.

- Needling occurs at the surface of nonwoven cloth/fiber web, and during the process of needling, the composite moves horizontally with the conveyor belt, and the needle plate moves up and down at a certain frequency.

- The composite is rotated horizontally for 90°, and the needling process is repeated to ensure the uniformity of the needle holes in the X and Y directions.

Repeat the preceding three steps until the preform reaches a certain thickness and needling density. In the process of needling, part of the in-plane fiber was introduced into the ply thickness direction to generate vertical fiber clusters, so that the carbon fiber and short-chopped-fiber web are closely combined to form a preform with a certain strength in the plane and between layers.

Table 7.1 shows composite fabric raw material and structure parameters. HTS™ (Toho, Tokyo, Japan) carbon fiber was used in twill woven cloth and T-700™ (Toray, Tokyo, Japan) carbon fiber was used in

TABLE 7.1   Composite Fabric Raw Materials and Woven Structural Parameters

Num	Fabric Preform	Fabric Forming Method	Density of Original Fabric / (kg/m³)	Fiber Volume of Original Fabric /(%)	Density of Heat-Treated Fabric / (kg/m³)	Fiber Volume of Heat-Treated Fabric /(%)
1#	HTS™-3K twill woven cloth/ T-700™-12K [±45°] plain-woven cloth/T-700™-12K short-chopped-fiber web	Needle	540	30	460	25.6
2#	HTS™-3K twill woven ply/ T-700™-12K [0°] nonwoven cloth/ T-700™-12K [±45°] plain-woven cloth/ T-700™-12K short-chopped-fiber web	Needle	670	37.2	590	32.8
3#	Two plies of HTS™-3K twill woven cloth/ T-700™ [±45°] plain-woven cloth/T-700™-12K short-chopped-fiber web	Needle	680	37.8	470	26.1
4#	Two plies of HTS™-3K twill woven cloth/ T-700™-12K [±45°] plain-woven cloth/ T-700™-12K short-chopped-fiber web	Needle + Stitch	670	37.2	580	32.2

plain-woven cloth and short-chopped-fiber web. For HTS™ carbon fiber, the fiber strength is $\sigma_{fc} = 4.2$ GPa, fiber modulus is $E_f = 240$ GPa, fracture strain is $\varepsilon_f = 1.8\%$, and the density is $d_f = 1790$ kg/m³. For T-700™ carbon fiber, the fiber strength is $\sigma_{fc} = 4.9$ GPa, fiber modulus is $E_f = 230$ GPa, fracture strain is $\varepsilon_f = 2.1\%$, and the density is $d_f = 1790$ kg/m³. Four different types of fabric preform were introduced for fabricating 3D needled-punched C/SiC composites:

- Type 1, the fabric preform is formed using the needle method, and is composed of three layers, including (1) HTS™-3K twill woven cloth, (2) T-700™-12K [±45°] plain-woven cloth, and (3) T-700™-12K short-chopped-fiber web.

- Type 2, the fabric preform is formed using the needle method and composed of four layers, including (1) HTS™-3K twill woven ply, (2) T-700™-12K [0°] nonwoven cloth, and (3) T-700™-12K [±45°] plain-woven cloth; (4) T-700™-12K short-chopped-fiber web.

- Type 3, the fabric preform is formed using the needle method and composed of four layers, including (1) two layers of HTS™-3K twill woven cloth, (2) T-700™ [±45°] plain-woven cloth, and (3) T-700™-12K short-chopped-fiber web.

- Type 4, the fabric preform is formed using the needle-and-stitch method and composed of four layers, including (1) two layers of HTS™-3K twill woven cloth, (2) T-700™-12K [±45°] plain-woven cloth, and (3) T-700™-12K short-chopped-fiber web.

In order to improve the surface performance of carbon fibers and release the residual stress of the preform, the needled-punch preform was heat-treated at elevated temperatures $T = 1800\text{--}2000°C$ in Ar atmosphere. The density of four original fabric preform was in the range of 540–680 kg/m³, and after heat-treated, the density of four fabric preform decreased and lay in the range of 460–580 kg/m³. The fiber volumes of four original fabric preform were in the range $V_f = 30\text{--}37.2\%$, and after being heat-treated, the fiber volumes of four fabric preform decreased and lay in the range $V_f = 25.6\text{--}32.8\%$. The fiber volume was calculated by

$$V_f = \frac{m_{preform}}{V_{preform} \times d_f}, \tag{7.1}$$

where $m_{preform}$ is the preform mass, $V_{preform}$ is the volume of preform, and d_f is the fiber density.

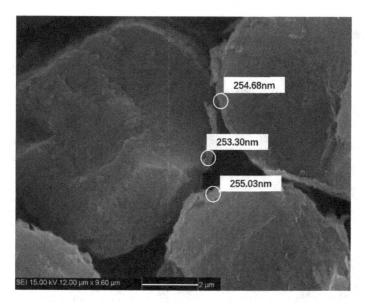

FIGURE 7.1   SEM micrograph of pyrolytic carbon interface layer on carbon fiber.

PyC was deposited on the carbon fiber surface as the interphase by the CVD process at $T = 850\,°C$ for $t = 20-50$ h. The PyC interphase thickness was approximately 200–300 nm, as shown in Figure 7.1. Using propylene and natural gas as precursors and nitrogen as a carrier-diluting gas, the CVD process was carried out for approximately $t = 200-300$ h to form the porous C/C composites. The density of porous C/C composites was approximately 1400–1500 kg/m³. As the carbon matrix is introduced many times, the carbon matrix was still coated on the surface of carbon fiber monofilament under the low-density state of C/C porous composite, forming a state similar to multilayer carbon interphase, as shown in Figures 7.2a and 7.2b. With the increase of carbon matrix, the subsequent PyC was coated outside the fiber bundle, as shown in Figure 7.2c and 7.2d. C/SiC composites were prepared by reactive infiltration of Si powder into C/C porous composite. The molten silicon reacted with the carbon matrix to form SiC matrix. C/SiC composite was prepared by coexistence of C matrix and SiC matrix. Figure 7.3 shows the macroscopic morphology of C/SiC composite after reactive infiltration.

The dog-bone-shaped specimens, with dimensions of 130 mm in length, 5 mm thick, and 12 mm wide in the gauge section, were cut

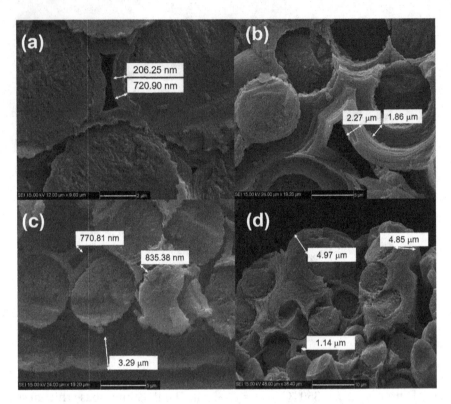

FIGURE 7.2 (a) First CVD carbon matrix, (b) second CVD carbon matrix, (c) multiple CVD carbon matrix, and (d) multiple CVD carbon matrix.

from 300 mm × 300 mm panels using wire-electrode cutting. Figure 7.4 shows the specimen size and configurations based on DqES415-2005 standard [17] with 30 mm in the testing gauge length. Monotonic and cyclic loading/unloading tensile tests of 3D needle-punched C/SiC composites were conducted on a SANS CMT5105 testing machine (MTS Systems Corp., Minneapolis, MN, USA) at room temperature. Monotonic and cyclic loading/unloading tensile tests were conducted under displacement control. A clip-on extensimeter was used to obtain the composite strain under monotonic and cyclic loading/unloading strain, as shown in Figure 7.5. The crosshead speed was 2.0 mm/min for monotonic tensile tests, and 0.5 mm/min for cyclic loading/unloading tensile tests. To analyze failure mechanisms of the composites, the microstructures of the fracture surfaces of the specimens were observed by FEI Quanta 200 field emission environmental scanning electron microscopy.

FIGURE 7.3 Macroscopic morphology of C/SiC composite after reactive infiltration.

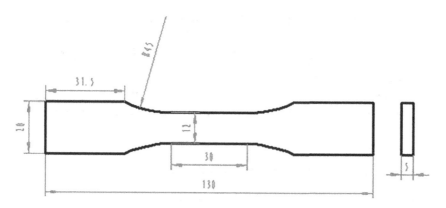

FIGURE 7.4 Specimen configuration for monotonic and loading/unloading cyclic tension.

The accelerating voltage is 15 and 20 kV. The vacuum value is set at 3×10^{-3} Pa. However, due to the effect of environmental temperature, the vacuum may change a little during the process of scanning electron microscopy (SEM) measurements.

FIGURE 7.5   Photograph of the tensile specimen and the clip-on extensometer.

7.3 MICROMECHANICAL HYSTERESIS CONSTITUTIVE MODEL

Under cyclic loading/unloading tensile, hysteresis loops would appear due to the interface slip in different damage regions. Based on the damage mechanisms of the interface debonding and slip, the constitutive model of hysteresis loops can be divided into two cases:

- Case 1, the interface partial debonding

- Case 2, the interface complete debonding

7.3.1 Interface Partial Debonding

For Case 1 with the partial debonding at the fiber interface, the unloading and reloading of the composite's strain are

$$\varepsilon_{\text{unloading}} = \frac{\Phi_U}{E_f} + 4\frac{\tau_i}{E_f}\frac{L_y^{\,2}}{r_f L_c} - \frac{\tau_i}{E_f}\frac{\left(2L_y - L_d\right)\left(2L_y + L_d - L_c\right)}{r_f L_c} - \left(\alpha_c - \alpha_f\right)\Delta T,$$

$$(7.2)$$

$$\varepsilon_{\text{reloading}} = \frac{\Phi_R}{E_f} - 4\frac{\tau_i}{E_f}\frac{L_z^2}{r_f L_c} + 4\frac{\tau_i}{E_f}\frac{\left(L_y - 2L_z\right)^2}{r_f L_c}$$

$$+ 2\frac{\tau_i}{E_f}\frac{\left(L_d - 2L_y + 2L_z\right)\left(L_d + 2L_y - 2L_z - L_c\right)}{r_f L_c} - \left(\alpha_c - \alpha_f\right)\Delta T,$$

(7.3)

where $\varepsilon_{\text{unloading}}$ and $\varepsilon_{\text{reloading}}$ denote the unloading and reloading composite's strain; Φ_U and Φ_R denote the intact fiber stress on unloading and reloading; τ_i is the interface shear stress in the debonding region; E_f is the fiber's elastic modulus; r_f is the fiber's radius; L_d is the interface debonding length, which can be determined using the fracture mechanic approach, as shown in Equation 7.4; L_y is the unloading interface counter-slip length, which can be determined using the fracture mechanics approach, as shown in Equation 7.5; and L_c is the matrix crack spacing, which can be determined using the stochastic matrix cracking model, as shown in Equation 7.6.

$$L_d = \frac{r_f}{2}\left(\frac{V_m E_m}{E_c \tau_i}\Phi - \frac{1}{\rho}\right)$$

$$- \sqrt{\left(\frac{r_f}{2\rho}\right)^2 - \frac{r_f^2 V_f V_m E_f E_m \Phi}{4E_c^2 \tau_i^2}\left(\Phi - \frac{\sigma}{V_f}\right) + \frac{r_f V_m E_m E_f}{E_c \tau_i^2}\Gamma_d},$$

(7.4)

$$L_y = \frac{1}{2}\left\{L_d - \left[\frac{r_f}{2}\left(\frac{V_m E_m}{E_c \tau_i}\Phi_U - \frac{1}{\rho}\right)\right.\right.$$

$$\left.\left. - \sqrt{\left(\frac{r_f}{2\rho}\right)^2 - \frac{r_f^2 V_f V_m E_f E_m \Phi_U}{4E_c^2 \tau_i^2}\left(\Phi_U - \frac{\sigma}{V_f}\right) + \frac{r_f V_m E_m E_f}{E_c \tau_i^2}\Gamma_d}\right]\right\},$$

(7.5)

$$L_c = L_{\text{sat}}\left\{1 - \exp\left[-\left(\frac{\sigma_m}{\sigma_R}\right)^m\right]\right\}^{-1},$$

(7.6)

where V_f and V_m denote the volume fraction of the fiber and the matrix, respectively; E_m and E_c denote the elastic modulus of the matrix and composite, respectively; ρ is the shear-lag model parameter; Γ_d is the interface debonding energy; L_{sat} is the saturation matrix crack spacing; σ_m is the stress in the matrix; σ_R is the characteristic cracking stress of the matrix; and m is the matrix Weibull modulus.

7.3.2 Interface Complete Debonding

For the complete debonding at the fiber interface, the unloading and reloading composite's strain is

$$\varepsilon_{\text{unloading}} = \frac{\Phi_U}{E_f} + 4\frac{\tau_i}{E_f}\frac{L_y^2}{r_f l_c} - 2\frac{\tau_i}{E_f}\frac{\left(2L_y - L_c/2\right)^2}{r_f L_c} - \left(\alpha_c - \alpha_f\right)\Delta T, \quad (7.7)$$

$$\varepsilon_{\text{reloading}} = \frac{\Phi_R}{E_f} - 4\frac{\tau_i}{E_f}\frac{L_z^2}{r_f L_c} + 4\frac{\tau_i}{E_f}\frac{\left(L_y - 2L_z\right)^2}{r_f L_c} - 2\frac{\tau_i}{E_f}\frac{\left(L_c/2 - 2L_y + 2L_z\right)^2}{r_f L_c}$$
$$- \left(\alpha_c - \alpha_f\right)\Delta T,$$

$$(7.8)$$

where

$$L_z = L_y - \frac{1}{2}\left\{ L_d - \left[\frac{r_f}{2}\left(\frac{V_m E_m}{E_c \tau_i}\Phi_R - \frac{1}{\rho} \right) \right. \right.$$
$$\left. \left. - \sqrt{\left(\frac{r_f}{2\rho} \right)^2 - \frac{r_f^2 V_f V_m E_f E_m \Phi_R}{4E_c^2 \tau_i^2}\left(\Phi_R - \frac{\sigma}{V_f} \right) + \frac{r_f V_m E_m E_f}{E_c \tau_i^2}\Gamma_d} \right] \right\}. \quad (7.9)$$

In the present analysis, the interface slip ratio (ISR) upon unloading and reloading is used to characterize the interface slip condition in the debonding regions:

$$\eta = \frac{2L_y}{L_c}, \gamma = \frac{2L_z}{L_c}. \quad (7.10)$$

7.4 EXPERIMENTAL COMPARISONS

In this section, the experimental cyclic loading/unloading hysteresis loops of Type 1–4 3D needle-punched C/SiC composites are predicted using the micromechanical hysteresis loops considering different interface debonding conditions.

7.4.1 Type 1 3D Needle-Punched C/SiC Composite

Figures 7.6 and 7.7 show the experimental and predicted cyclic loading/unloading hysteresis loops and interface slip ratio for different peak stresses of σ_{max} = 20, 30, 40, 50, and 60 MPa. With increasing peak stress from σ_{max} = 20–60 MPa, multiple damage mechanisms of matrix cracking, interface debonding, and fiber failure occur, leading to the evolution

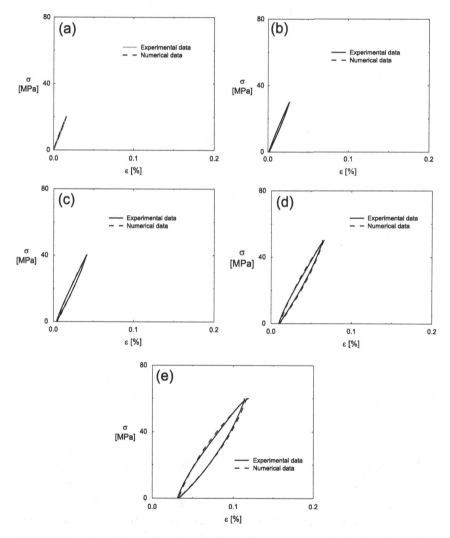

FIGURE 7.6 Experimental and predicted hysteresis loops of Type 1 3D needle-punched C/SiC composite under (a) σ_{max} = 20 MPa, (b) σ_{max} = 30 MPa, (c) σ_{max} = 40 MPa, (d) σ_{max} = 50 MPa, and (e) σ_{max} = 60 MPa.

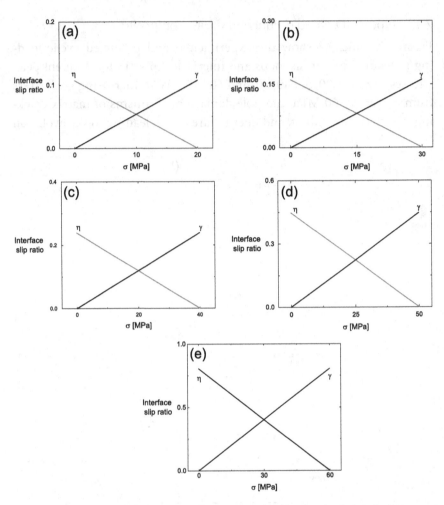

FIGURE 7.7 Interface slip ratio of Type 1 3D needle-punched C/SiC composite under (a) σ_{max} = 20 MPa, (b) σ_{max} = 30 MPa, (c) σ_{max} = 40 MPa, (d) σ_{max} = 50 MPa, and (e) σ_{max} = 60 MPa.

of the mechanical hysteresis appearance. The damage parameters of composite's residual strain ε_{res}, composite's peak strain ε_p, composite's hysteresis loops width $\Delta\varepsilon$, and the hysteresis loops area U, interface slip ratio η and γ, and the loading/unloading inverse tangent modulus (ITM) relate with composite's internal damages. The changing of these damage parameters with increasing peak stress can reflect the damage extent inside of the composite.

TABLE 7.2   Hysteresis Parameters of Type 1 3D Needle-Punched C/SiC Composite

Peak Stress/(MPa)	ε_{res}/(%)	ε_p/(%)	$\Delta\varepsilon$/(%)	ΔW/(kPa)	η_{max}/γ_{max}
20	0.0003	0.0158	0.0016	0.18	0.107
30	0.00136	0.0266	0.002	0.55	0.161
40	0.00435	0.0411	0.00369	1.38	0.238
50	0.01052	0.064	0.00715	3.6	0.447
60	0.03173	0.113	0.017	10.5	0.805

Table 7.2 listed the evolution of damage parameters with increasing peak stress. The composite's residual strain increased from ε_{res} = 0.0003 to 0.03% mainly due to damage mechanisms of matrix cracking and interface debonding; the composite's peak strain increased from ε_p = 0.0158 to 0.113% mainly due to the damage mechanisms of matrix cracking, interface debonding, and fiber's failure; the composite's hysteresis loops width increased from $\Delta\varepsilon$ = 0.0016 to 0.017% mainly due to the increase of the interface debonding and slip range in the composite; and the composite's hysteresis loops area increased from ΔW = 0.18 to 10.5 kPa mainly due to the decrease of matrix crack spacing and increase of the interface debonding and slip range. With increasing tensile peak stress from σ_{max} = 20 to 60 MPa, the interface debonding and slip range increased, and the interface slip ratio at the valley stress and the peak stress increased from $\eta(\sigma_{min})$ = $\gamma(\sigma_{max})$ = 0.107 to 0.805.

Figure 7.8 shows the composite's ITM on unloading and reloading versus the strain curves. For the unloading and reloading ITM, the composite's ITM increased with decreasing composite's strain, and the curve can be divided into two stages, that is, (1) on initial unloading or reloading, the composite's unloading or reloading ITM increased rapidly with decreasing or increasing strain due to the increasing interface debonding and slip range, and (2) the unloading or reloading ITM increased slowly with decreasing or increasing strain as the interface slip approaching the interface debonding tip.

- On unloading at the peak stress σ_{max} = 20 MPa, the composite's unloading ITM increased rapidly from Σ_{ITM} = 2.3 TPa^{-1} to Σ_{ITM} = 6.05 TPa^{-1} due to the increase of the interface slip range and then increased slowly to Σ_{ITM} = 7.9 TPa^{-1} as the interface slip range approaches the interface debonding tip; on reloading from the valley stress σ_{min} = zero MPa, the composite's reloading ITM increased rapidly from

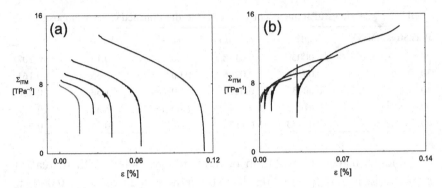

FIGURE 7.8 Evolution of Type 1 C/SiC composite's ITM versus strain for (a) unloading and (b) reloading.

$\Sigma_{ITM} = 3.2$ TPa^{-1} to $\Sigma_{ITM} = 6.1$ TPa^{-1} due to the increase of the interface slip range and then increased slowly to $\Sigma_{ITM} = 7.9$ TPa^{-1} as the interface slip range approaches the interface debonding tip.

- On unloading at peak stress $\sigma_{max} = 60$ MPa, the composite's unloading ITM increased rapidly from $\Sigma_{ITM} = 0.27$ TPa^{-1} to $\Sigma_{ITM} = 6.69$ TPa^{-1} due to the increase of the interface slip range, and then increased slowly to $\Sigma_{ITM} = 13.6$ TPa^{-1} as the interface slip range approaches the interface debonding tip; upon reloading from the valley stress $\sigma_{min} = $ zero MPa, the composite's reloading ITM increased rapidly from $\Sigma_{ITM} = 4$ TPa^{-1} to $\Sigma_{ITM} = 7.5$ TPa^{-1} due to the increase of the interface slip range, and then increased slowly to $\Sigma_{ITM} = 14.5$ TPa^{-1} as the interface slip range approaches the interface debonding tip.

For the Type 1 3D needle-punched C/SiC composite, the experimental cyclic loading/unloading hysteresis loops under the peak stresses $\sigma_{max} = $ 20, 30, 40, 50, and 60 MPa all correspond to the interface debonding state Case 1. Upon unloading and reloading, the interface partial debonds and the fiber slides partial relative to the matrix in the interface debonding region. The theoretical predicted cyclic loading/unloading hysteresis loops agreed with experimental data.

7.4.2 Type 2 3D Needle-Punched C/SiC Composite

Figures 7.9 and 7.10 show the experimental and predicted cyclic loading/unloading tensile hysteresis loops and interface slip ratio for different peak stresses $\sigma_{max} = $ 40, 50, 60, 70, 80, 90, and 100 MPa. Table 7.3 listed

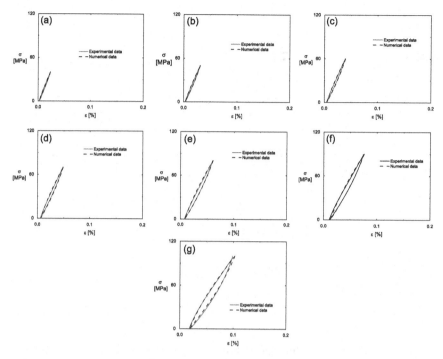

FIGURE 7.9   Experimental and predicted hysteresis loops of Type 2 3D needle-punched C/SiC composite under (a) σ_{max} = 40 MPa, (b) σ_{max} = 50 MPa, (c) σ_{max} = 60 MPa, (d) σ_{max} = 70 MPa, (e) σ_{max} = 80 MPa, (f) σ_{max} = 90 MPa, and (g) σ_{max} = 100 MPa.

the evolution of damage parameters with increasing peak stress. With increasing tensile peak stress from σ_{max} = 40 to 100 MPa, the composite's residual strain increased from ε_{res} = 0.002 to 0.018% mainly attributed to the damage mechanisms of matrix cracking and interface debonding; the composite's peak strain increased from ε_p = 0.023 to 0.097%, mainly attributed to the damage mechanisms of matrix cracking, interface debonding, and fiber failure; the composite's hysteresis loops width increased from $\Delta\varepsilon$ = 0.0019 to 0.0139%, mainly due to the increased range of the interface debonding and slip; and the composite's hysteresis loops area increased from ΔW = 0.549 to 14.7 kPa mainly due to the decrease of matrix crack spacing and increase of the interface debonding and slip range. With increasing tensile peak stress from σ_{max} = 40 to 100 MPa, the interface slip ratio at the valley stress and peak stress increased from $\eta(\sigma_{min}) = \gamma(\sigma_{max})$ = 0.11 to 0.339 as the interface slip approaching the interface debonding tip.

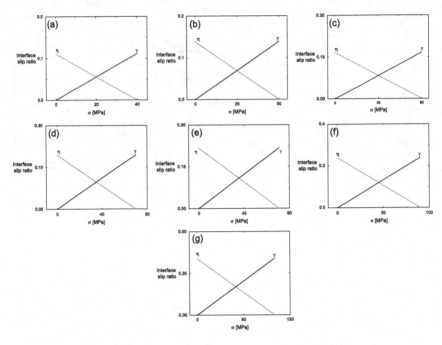

FIGURE 7.10   Interface slip ratio of Type 2 3D needle-punched C/SiC composite under (a) σ_{max} = 40 MPa, (b) σ_{max} = 50 MPa, (c) σ_{max} = 60 MPa, (d) σ_{max} = 70 MPa, (e) σ_{max} = 80 MPa, (f) σ_{max} = 90 MPa, and (g) σ_{max} = 100 MPa.

TABLE 7.3   Hysteresis Parameters of Type 2 3D Needle-Punched C/SiC Composite

Peak Stress/(MPa)	ε_{res}/(%)	ε_p/(%)	$\Delta\varepsilon$/(%)	ΔW/(kPa)	η_{max}/γ_{max}
40	0.00213	0.0233	0.0019	0.549	0.110
50	0.0032	0.0304	0.0023	0.867	0.138
60	0.0045	0.0387	0.003	1.4	0.166
70	0.0059	0.048	0.004	2.2	0.193
80	0.0079	0.0595	0.005	3.8	0.217
90	0.011	0.0744	0.008	7.1	0.238
100	0.018	0.097	0.0139	14.7	0.339

Figure 7.11 shows the composite's ITM on unloading and reloading versus the strain curves. For the unloading and reloading ITM, the composite's ITM increased with decreasing composite's strain, and the curve can be divided into two stages, that is, (1) on initial unloading or reloading, the composite's unloading or reloading ITM increased rapidly with decreasing or increasing strain due to the increasing interface debonding and slip

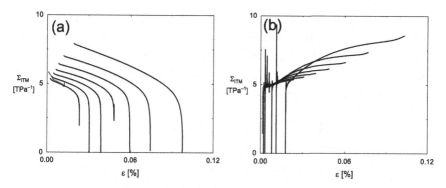

FIGURE 7.11 Evolution of Type 2 C/SiC composite's ITM versus strain for (a) unloading and (b) reloading.

range, and (2) the unloading or reloading ITM increased slowly with decreasing or increasing strain as the interface slip range approaches the interface debonding tip.

- On unloading at peak stress σ_{max} = 40 MPa, the composite's unloading ITM increased rapidly from Σ_{ITM} = 1.95 TPa^{-1} to Σ_{ITM} = 4.09 TPa^{-1} due to the increasing interface debonding and slip range and then increased slowly to Σ_{ITM} = 5.3 TPa^{-1} as the interface slip range approaches the interface debonding tip; on reloading from the valley stress, the composite's reloading ITM increased rapidly from Σ_{ITM} = 2.3 TPa^{-1} to Σ_{ITM} = 4.75 TPa^{-1} due to the increasing interface debonding and slip range and then increased slowly to Σ_{ITM} = 5.38 TPa^{-1} as the interface slip range approaches the interface debonding tip.

- On unloading at peak stress σ_{max} = 100 MPa, the composite's unloading ITM increased rapidly from Σ_{ITM} = 0.05 TPa^{-1} to Σ_{ITM} = 4.1 TPa^{-1} due to the increasing interface debonding and slip range and then increased slowly to Σ_{ITM} = 7.9 TPa^{-1} as the interface slip range approaches the interface debonding tip; on reloading from the valley stress, the composite's reloading ITM increased rapidly from Σ_{ITM} = 1.4 TPa^{-1} to Σ_{ITM} = 4.9 TPa^{-1} due to the increasing interface debonding and slip range and then increased slowly to Σ_{ITM} = 8.5 TPa^{-1} as the interface slip range approaches the interface debonding tip.

For Type 2 3D needle-punched C/SiC composite, the experimental cyclic loading/unloading hysteresis loops under different peak stresses

σ_{max} = 40, 50, 60, 70, 80, 90, and 100 MPa correspond to the interface debonding state of Case 1. On unloading and reloading, the interface partially debonds and the fiber slides partially relative to the matrix in the interface debonding region. The theoretical predicted cyclic loading/unloading hysteresis loops agreed with experimental data.

7.4.3 Type 3 3D Needle-Punched C/SiC Composite

Figures 7.12 and 7.13 show the experimental and predicted cyclic loading/unloading hysteresis loops and interface slip ratio for different peak stresses σ_{max} = 20, 30, 40, 50, and 60 MPa. Table 7.4 lists the evolution of damage parameters with increasing peak stress. With increasing tensile peak stress from σ_{max} = 20 to 60 MPa, the composite's residual strain increased from ε_{res} = 0.001 to 0.019% mainly due to the damage mechanisms of matrix cracking and interface debonding; the composite's peak strain increased from ε_p = 0.017 to 0.101% mainly attributed to the damage mechanisms of matrix cracking, interface debonding, and fiber failure; the composite's hysteresis loops width increased from $\Delta\varepsilon$ = 0.0013 to 0.0144% mainly attributed to the increase of interface debonding and slip range; and the composite's hysteresis loops area increased from ΔW = 0.196 to 9.0 kPa mainly attributed to the decrease of the matrix crack spacing and the increase of the interface debonding and slip range. With increasing tensile peak stress from σ_{max} = 20 to 60 MPa, the interface slip ratio increases from $\eta(\sigma_{min}) = \gamma(\sigma_{max})$ = 0.146 to 0.729.

Figure 7.14 shows the composite's ITM on unloading and reloading versus the strain curves. For the unloading and reloading ITM, the composite's ITM increased with decreasing composite's strain, and the curve can be divided into two stages, that is, (1) on initial unloading or reloading, the composite's unloading or reloading ITM increased rapidly with decreasing or increasing strain due to the increase in the interface slip range, and (2) the unloading or reloading ITM increased slowly with decreasing or increasing strain as the interface slip range approaches the interface debonding tip.

- On unloading at peak stress σ_{max} = 20 MPa, the composite's unloading ITM increased rapidly from Σ_{ITM} = 5.17 TPa^{-1} to Σ_{ITM} = 6.19 TPa^{-1} due to the increase in the interface slip range and then increased slowly to Σ_{ITM} = 8.2 TPa^{-1} as the interface slip range approaches the interface debonding tip; on reloading from the valley stress, the

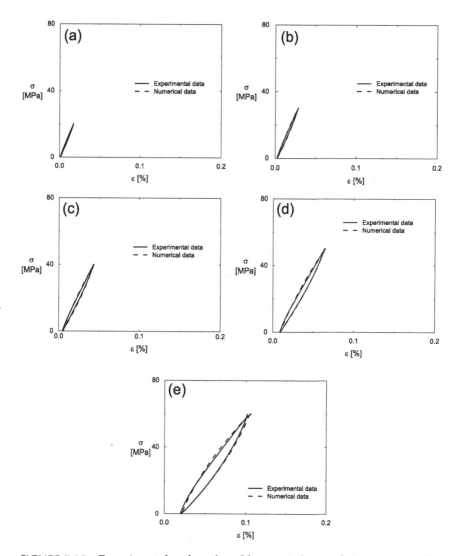

FIGURE 7.12   Experimental and predicted hysteresis loops of Type 3 3D needle-punched C/SiC composite under (a) σ_{max} = 20 MPa, (b) σ_{max} = 30 MPa, (c) σ_{max} = 40 MPa, (d) σ_{max} = 50 MPa, and (e) σ_{max} = 60 MPa.

composite's reloading ITM increased rapidly from Σ_{ITM} = 1.1 TPa^{-1} to Σ_{ITM} = 6.1 TPa^{-1} due to the increase in the interface slip range and then increased slowly to Σ_{ITM} = 8.22 TPa^{-1} as the interface slip range approaches the interface debonding tip.

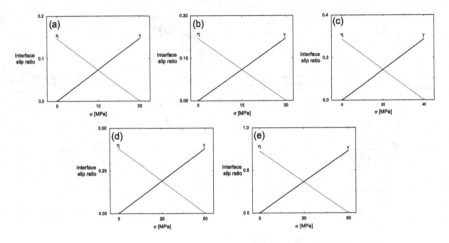

FIGURE 7.13 Interface slip ratio of Type 3 3D needle-punched C/SiC composite under (a) σ_{max} = 20 MPa, (b) σ_{max} = 30 MPa, (c) σ_{max} = 40 MPa, (d) σ_{max} = 50 MPa, and (e) σ_{max} = 60 MPa.

TABLE 7.4   Hysteresis Parameters of Type 3 3D Needle-Punched C/SiC Composite

Peak stress/(MPa)	$\varepsilon_{res}/(\%)$	$\varepsilon_p/(\%)$	$\Delta\varepsilon/(\%)$	$\Delta W/(kPa)$	η_{max}/γ_{max}
20	0.00103	0.017	0.00133	0.196	0.146
30	0.00247	0.028	0.00215	0.539	0.219
40	0.00484	0.042	0.003	1.21	0.286
50	0.00893	0.063	0.0062	3.09	0.379
60	0.0199	0.101	0.0144	9	0.729

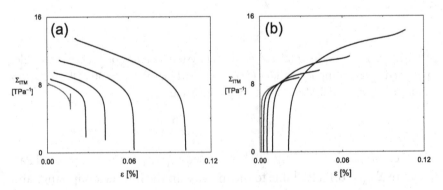

FIGURE 7.14 Evolution of Type 3 C/SiC composite's ITM versus strain for (a) unloading and (b) reloading.

- On unloading at peak stress σ_{max} = 60 MPa, the composite's unloading ITM increased rapidly from Σ_{ITM} = 0.28 TPa^{-1} to Σ_{ITM} = 5.97 TPa^{-1} due to the increase in the interface slip range, and then increased slowly to Σ_{ITM} = 13.4 TPa^{-1} as the interface slip range approaches the interface debonding tip; on reloading from the valley stress, the composite's reloading ITM increased rapidly from Σ_{ITM} = 0.16 TPa^{-1} to Σ_{ITM} = 6.93 TPa^{-1} due to the increase in the interface slip range and then increased slowly to Σ_{ITM} = 14.3 TPa^{-1} as the interface slip range approaches the interface debonding tip.

For Type 3 3D needle-punched C/SiC composite, the experimental cyclic loading/unloading hysteresis loops under different peak stresses σ_{max} = 20, 30, 40, 50, and 60 MPa all correspond to the interface debonding state of Case 1. On unloading and reloading, the interface partially debonds, and the fiber slides partially relative to the matrix in the interface debonding region. The theoretical predicted cyclic loading/unloading hysteresis loops agreed with experimental data.

7.4.4 Type 4 3D Needle-Punched C/SiC Composite

Figures 7.15 and 7.16 show the experimental and predicted cyclic loading/unloading hysteresis loops and interface slip ratio for different tensile peak stresses σ_{max} = 30, 40, 50, 60, 70, 80, and 90 MPa. Table 7.5 lists the evolution of damage parameters with increasing peak stress. With increasing tensile peak stress from σ_{max} = 30 to 90 MPa, the composite's residual strain increased from ε_{res} = 0.002 to 0.035% mainly attributed to the damage mechanisms of matrix cracking and interface debonding; the composite's peak strain increased from ε_p = 0.0228 to 0.148% mainly attributed to the damage mechanisms of matrix cracking, interface debonding, and fiber failure; the composite's hysteresis loops width increased from $\Delta\varepsilon$ = 0.0013 to 0.021% mainly attributed to the increased range of the interface debonding and slip; and the composite's hysteresis loops area increases from ΔW = 0.418 to 21.5 kPa mainly attributed to the decrease of the matrix crack spacing and the increase of the interface debonding and slip range. With increasing tensile peak stress from σ_{max} = 30 to 90 MPa, the interface slip ratio increased from $\eta(\sigma_{min})$ = $\gamma(\sigma_{max})$ = 0.102 to 0.619.

Figure 7.17 shows the composite's ITM on unloading and reloading versus the strain curves. For the unloading and reloading ITM, the composite's ITM increased with decreasing composite's strain, and the curve can

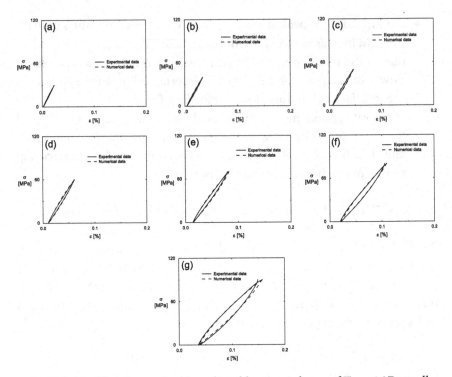

FIGURE 7.15 Experimental and predicted hysteresis loops of Type 4 3D needle-punched C/SiC composite under (a) σ_{max} = 30 MPa, (b) σ_{max} = 40 MPa, (c) σ_{max} = 50 MPa, (d) σ_{max} = 60 MPa, (e) σ_{max} = 70 MPa, (f) σ_{max} = 80 MPa, and (g) σ_{max} = 90 MPa.

be divided into two stages, that is, (1) on initial unloading or reloading, the composite's unloading or reloading ITM increased rapidly with decreasing or increasing strain due to the increase of the interface slip range, and (2) the unloading or reloading ITM increased slowly with decreasing or increasing strain as the interface slip range approaches the interface debonding tip.

- On unloading at peak stress σ_{max} = 30 MPa, the composite's unloading ITM increased rapidly from Σ_{ITM} = 0.58 TPa^{-1} to Σ_{ITM} = 5.21 TPa^{-1} due to the increase of the interface slip range, and then increased slowly to Σ_{ITM} = 6.96 TPa^{-1} as the interface slip range approaches the interface debonding tip; on reloading from the valley stress, the composite's reloading ITM increased rapidly from Σ_{ITM} = 2.38 TPa^{-1}

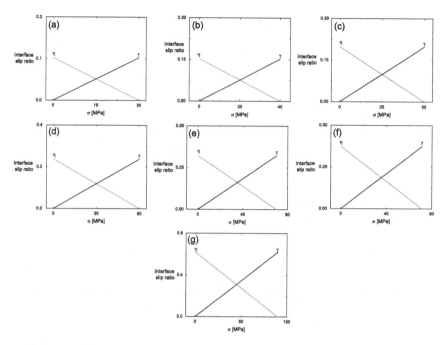

FIGURE 7.16   Interface slip ratio of Type 4 3D needle-punched C/SiC composite under (a) σ_{max} = 30 MPa, (b) σ_{max} = 40 MPa, (c) σ_{max} = 50 MPa, (d) σ_{max} = 60 MPa, (e) σ_{max} = 70 MPa, (f) σ_{max} = 80 MPa, and (g) σ_{max} = 90 MPa.

TABLE 7.5   Hysteresis Parameters of Type 4 3D Needle-Punched C/SiC Composite

Peak Stress/(MPa)	ε_{res}/(%)	ε_p/(%)	$\Delta\varepsilon$/(%)	ΔW/(kPa)	η_{max}/γ_{max}
30	0.002	0.0228	0.0013	0.418	0.102
40	0.004	0.033	0.0019	0.752	0.152
50	0.0067	0.045	0.003	1.36	0.198
60	0.0098	0.065	0.0041	2.5	0.237
70	0.014	0.079	0.007	4.8	0.322
80	0.019	0.105	0.012	9.9	0.373
90	0.035	0.148	0.021	21.5	0.619

to Σ_{ITM} = 5.9 TPa^{-1} due to the increase of the interface slip range, and then increased slowly to Σ_{ITM} = 7.2 TPa^{-1} as the interface slip range approaches the interface debonding tip.

- On unloading at peak stress σ_{max} = 90 MPa, the composite's unloading ITM increased rapidly from Σ_{ITM} = 0.05 TPa^{-1} to Σ_{ITM} = 5.42 TPa^{-1}

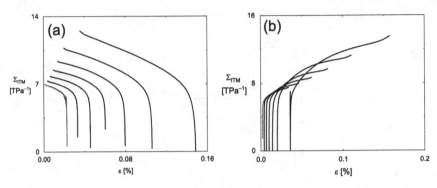

FIGURE 7.17 Evolution of Type 4 C/SiC composite's ITM versus strain for (a) unloading and (b) reloading.

due to the increase of the interface slip range and then increased slowly to Σ_{ITM} = 12.5 TPa^{-1} as the interface slip range approaches the interface debonding tip; on reloading from the valley stress, the composite's reloading ITM increased rapidly from Σ_{ITM} = 1.43 TPa^{-1} to Σ_{ITM} = 6.5 TPa^{-1} due to the increase of the interface slip range and then increased slowly to Σ_{ITM} = 13.5 TPa^{-1} as the interface slip range approaches the interface debonding tip.

For the Type 4 3D needle-punched C/SiC composite, the experimental cyclic loading/unloading hysteresis loops under different peak stresses σ_{max} = 30, 40, 50, 60, 70, 80, and 90 MPa all correspond to the interface debonding state of Case 1. On unloading and reloading, the interface partial debonds and the fiber slides partial relative to the matrix in the interface debonding region. The theoretical predicted cyclic loading/unloading hysteresis loops agreed with experimental data.

7.5 SUMMARY AND CONCLUSION

In this chapter, cyclic loading/unloading tensile hysteresis behavior of 3D needle-punched C/SiC composite was investigated using a micromechanical hysteresis constitutive relationship model considering damage mechanisms of matrix cracking, interface debonding and slip, and fiber fracture and pullout. Hysteresis loops of four different types of 3D needle-punched C/SiC composites were predicted for different peak stresses. Hysteresis parameters of unloading residual strain, peak strain, hysteresis loops width, hysteresis loops area, interface slip ratio, and ITM were adopted to characterize the tensile damage evolution inside of composites.

- Experimental cyclic loading/unloading curves of different 3D needle-punched C/SiC composites show obvious hysteresis loops. The composite's residual strain, peak strain, hysteresis width, and hysteresis area increased with peak stress, indicating damage evolution and propagation inside of composites.

- With increasing peak stress, the interface slip ratio of counter-slip ratio and new-slip ratio increased with peak stress, indicating the increase of the interface debonding and slip range inside of the composites.

- Predicted hysteresis loops of different 3D needle-punched C/SiC composites agreed with experimental data.

REFERENCES

1. Li L.B. *Durability of ceramic matrix composites.* Woodhead Publishing, Oxford, UK. 2020.
2. Chen X, Chen L, Zhang C, Song L, Zhang D. Three-dimensional needle-punching for composites – A review. *Compos. Part A* 2016; 85:12–30.
3. Nie J, Xu Y, Zhang L, Cheng L, Ma J. Microstructure and tensile behavior oof multiply needled C/SiC composite fabricated by chemical vapor infiltration. *J. Mater. Proc. Technol.* 2009; 209:572–576.
4. Fang P, Cheng L, Zhang L, Nie J. Monotonic tensile behavior analysis of three-dimensional needle-punched woven C/SiC composites by acoustic emission. *J. Univ. Sci. Technol.* 2008; 15:302–306.
5. Fan S, Zhang L, Cheng L, Xu F. Microstructure and compressive behavior of 3D needled C/SiC composites. *Adv. Mater. Res.* 2011; 194–196:1599–1606.
6. Chen Z, Fang G, Xie J, Liang J. Experimental study of high-temperature tensile mechanical properties of 3D needled C/C-SiC composites. *Mater. Sci. Eng. A* 2016; 654:271–277.
7. Nie J, Xu Y, Wan Y, Zhang L, Cheng L, Ma J. Tensile behavior of three-dimensional needed carbon fiber reinforced SiC composite under load-unload. *J. Chin. Ceram. Soc.* 2009; 37:76–82.
8. Liu YF, Li LB, Zhang ZW, Xiong X. Monotonic and cyclic loading/unloading tensile behavior of 3D needle-punched C/SiC ceramic-matrix composites. *Materials* 2021; 14:57.
9. Mei H, Cheng LF. Comparison of the mechanical hysteresis of carbon/ceramic-matrix composites with different fiber preforms. *Carbon* 2009; 47:1034–1042.
10. Li LB. Cyclic loading/unloading hysteresis behavior of fiber-reinforced ceramic-matrix composites at room and elevated temperatures. *Mater. Sci. Eng. A* 2015; 648:235–242.

11. Guo XJ, Wu JW, Li J, Zeng YQ, Huang XZ, Li LB. Damage monitoring of 2D SiC/SiC composites under monotonic and cyclic loading/unloading using acoustic emission and natural frequency. *Ceramics-Silikaty* 2021; Online. doi:10.13168/cs.2021.0011

12. Xie J, Fang G, Chen Z, Liang J. Shear nonlinear constitutive relationship of needled C/C-SiC composite. *Acta Mater. Compos. Sinica* 2016; 33:1507–1514.

13. Li LB., Song Y, Sun Y. Modeling the tensile behavior of unidirectional C/SiC ceramic-matrix composites. *Mech. Compos. Mater.* 2014; 49:659–672.

14. Li LB. Modeling the monotonic and cyclic tensile stress-strain behavior of 2D and 2.5D woven C/SiC ceramic-matrix composites. *Mech. Compos. Mater.* 2018; 54:165–178.

15. Li LB. Effect of stochastic loading on tensile damage and fracture of fiber-reinforced ceramic-matrix composites. *Mater* 2020; 13:2469.

16. Li LB. A micromechanical tension-tension fatigue hysteresis loops model of fiber-reinforced ceramic-matrix composites considering stochastic matrix fragmentation. *Int. J. Fatigue* 2020; 143:106001.

17. Aerospace Research Institute of Materials and Processing Technology. Test method for tensile properties of fine weave pierced carbon/carbon composites: DqES415-2005. Beijing: Aerospace Research Institute of Materials and Processing Technology, 2005.

CHAPTER 8

Mechanical Hysteresis Behavior in CMCs under Multiple-Stage Loading

8.1 INTRODUCTION

Under cyclic fatigue loading, the material performance of fiber-reinforced CMCs degrades with applied cycles due to the damage mechanisms of matrix multi-cracking, fiber/matrix interface debonding, and interface wear [1–7]. The nondestructive testing (NDT) techniques used for damage monitoring in structural metallic, that is, optical microscopy, acoustic emission (AE), acousto-ultrasonic, ultrasonic C-scan, X-ray, thermography, eddy current, and digital image correlation (DIC) have been adapted to characterize the fiber-reinforced CMCs at room or elevated temperature [8, 9]. Momon et al. [10] developed an AE-based technique to predict the residual fatigue life of two different types of·CMCs, that is, SiC/[Si-B-C] and C/[Si-B-C], under static fatigue at elevated temperature. The elastic energy released during static fatigue loading has been related to the AE activity to anticipate the final fracture of the specimen. Whitlow et al. [11] *in situ* monitored the localized damage and full-field surface strain of 2D SiC/SiC composite subjected to monotonic tensile loading using AE and DIC methods. The initiation and propagation of damage relate to the strain field changing with increasing applied stress. Simon et al. [12] developed an electrical resistance (ER)–based damage monitoring method for

DOI: 10.1201/b23026-8

SiC/[Si–B–C] composite under static and cyclic fatigue at 450°C in an air atmosphere. It was found that the ER is an accurate indicator of the damage stage of the interphase in the oxidizing environment. The application of these NDT methods mentioned earlier is based on the understanding of the macro-mechanical behavior and microstructural damages inside o fiber-reinforced CMCs. The stress-strain hysteresis loops are another way to monitor the internal damage accumulation of fiber-reinforced CMCs. The damage mechanisms of matrix multi-cracking, fiber/matrix interface debonding/sliding, and fiber fracture can be reflected on the fatigue hysteresis modulus, hysteresis width, and hysteresis energy dissipation [13–16]. Campbell and Jenkins [17] used the hysteresis modulus degradation and hysteresis energy dissipation to investigate the thermal degradation of an oxide/oxide CMC. Fantozzi and Reynaud [18] investigated the hysteresis behavior of bi- or multidirectional (cross–weave, cross-ply, 2.5D, $[0/+60/–60]_n$) with SiC or C long fiber–reinforced SiC, MAS-L, Si-B-C, or C matrix at room and elevated temperatures in inert and oxidation conditions. Li [19, 20] investigated the effects of fiber/matrix interface debonding, fibers oxidation, and fracture on matrix cracking in fiber-reinforced CMCs at room and elevated temperatures. Li [21] analyzed the strain response of fiber-reinforced CMCs subjected to stress-rupture and cyclic loading at elevated temperatures and compared the damage evolution of 2D C/SiC and SiC/SiC composites subjected to cyclic fatigue loading at room and elevated temperatures [22]. Li [23–28] analyzed the fiber/matrix interface debonding and sliding of fiber-reinforced CMCs under two-stage cyclic loading and developed the fatigue hysteresis loops models for fiber-reinforced CMCs. The strain energy density can be used to analyze the damage inside of materials [25–27]. Li [29–31] developed a hysteresis dissipated energy-based approach to predict the damage evolution and lifetime of fiber-reinforced CMCs under cyclic fatigue loading. The damage evolution and lifetime of unidirectional CMCs have been predicted. Under multiple loading sequences, the combination of the fatigue loading sequence and the damage mechanisms of matrix cracking, fiber/matrix interface debonding and sliding, and interface wear affect the damage evolution process of CMCs [32].

In this chapter, the mechanical hysteresis behavior of fiber-reinforced CMCs under multiple-stage loading is investigated. Considering the combination effects of multiple loading sequences and fatigue damage mechanisms, fatigue hysteresis loops, fatigue hysteresis dissipated energy, and fatigue hysteresis modulus are obtained to monitor the damage evolution process. Experimental damage evolution of C/SiC and SiC/SiC composites subjected to multiple fatigue loading sequences are predicted. The effects of fiber volume fraction, matrix crack spacing, fatigue peak stress, and fatigue stress range on damage evolution of the composite are discussed.

8.2 MICROMECHANICAL HYSTERESIS CONSTITUTIVE MODEL

If matrix cracking and fiber/matrix interface debonding are present upon first loading to the fatigue peak stress σ_{max1}, the hysteresis loops may develop as a result of energy dissipation through the frictional slip between fibers and the matrix. Under cyclic loading, the interface shear stress degrades from the initial value τ_i to τ_f. With increasing fatigue peak stress from σ_{max1} to σ_{max2}, cracks may propagate along the interface. The original interface debonding length at σ_{max1} is defined to be ξ, and the new interface debonding length at σ_{max2} is defined to be L_d. Based on the interface debonding and sliding range between matrix crack spacing, the fatigue hysteresis loops under the multiple loading sequence can be divided into four different cases:

- Case 1, the interface partially debonding (i.e., $L_d < L_c/2$), and the fiber sliding completely relative to the matrix in the interface debonding region (i.e., $L_y/L_d = L_z/L_d = 1$)

- Case 2, the interface partially debonding (i.e., $L_d < L_c/2$), and the fiber sliding partially relative to the matrix in the interface debonding region (i.e., $L_y/L_d = L_z/L_d < 1$)

- Case 3, the interface completely debonding (i.e., $L_d = L_c/2$), and the fiber sliding partially relative to the matrix in the interface debonding region (i.e., $L_y/L_d = L_z/L_d < 1$)

- Case 4, the interface completely debonding (i.e., $L_d = L_c/2$), and the fiber sliding completely relative to the matrix in the interface debonding region (i.e., $L_y/L_d = L_z/L_d = 1$).

8.2.1 Case 1

Upon unloading to σ ($\sigma_{min2} < \sigma < \sigma_{max2}$), the unit cell can be divided into the interface debonding region and the interface bonding region. The interface debonding region can be divided into three regions, that is, the interface counter-slip region with low interface shear stress τ_f ($x \in [0, \xi]$), interface slip region with high interface shear stress τ_i ($x \in [\xi, L_y]$), and interface slip region with high interface shear stress τ_i ($x \in [L_y, L_d]$). On unloading to the transition stress σ_{tr_pu} ($\sigma_{tr\_pu} > \sigma_{min2}$), the interface counter-slip length L_y approaches the interface debonding length L_d, that is, $L_y(\sigma_{tr\_pu}) = L_d$.

$$L_d = \left(1 - \frac{\tau_f}{\tau_i}\right)\xi + \frac{r_f}{2}\left(\frac{V_m E_m \sigma}{V_f E_c \tau_i} - \frac{1}{\rho}\right) - \sqrt{\left(\frac{r_f}{2\rho}\right)^2 + \frac{r_f V_m E_m E_f}{E_c \tau_i^2}\Gamma_i}, \quad (8.1)$$

$$L_y = \frac{1}{2}\left\{L_d\left(\sigma_{max2}\right) + \left(1 - \frac{\tau_f}{\tau_i}\right)\xi \right.$$
$$\left. - \left[\frac{r_f}{2}\left(\frac{V_m E_m}{V_f E_c}\frac{\sigma}{\tau_i} - \frac{1}{\rho}\right) - \sqrt{\left(\frac{r_f}{2\rho}\right)^2 + \frac{r_f V_m E_m E_f}{E_c \tau_i^2}\Gamma_i}\right]\right\}, \quad (8.2)$$

where r_f denotes the fiber radius; V_f and V_m denote the fiber and matrix volume fraction, respectively; E_f, E_m, and E_c denote the fiber, matrix, and composite elastic modulus, respectively; and Γ_i denotes the fiber/matrix interface debonded energy.

Upon reloading to σ ($\sigma_{min2} < \sigma < \sigma_{max2}$), slip again occurs near the matrix crack plane over a distance L_z, which is denoted to be the fiber/matrix interface new-slip region. The fiber/matrix interface debonding region can be divided into four regions, that is, interface new-slip region with low interface shear stress τ_f ($x \in [0, L_z]$), interface counter-slip region with low interface shear stress τ_f ($x \in [L_z, \xi]$), interface counter-slip region with high interface shear stress τ_i ($x \in [\xi, L_y]$), and interface slip region with high interface shear stress τ_i ($x \in [L_y, L_d]$). On reloading to the transition stress σ_{tr_pr} ($\sigma_{tr\_pr} < \sigma_{max2}$), the interface new-slip length L_z approaches the interface debonding length L_d, that is, $L_z(\sigma_{tr\_pr}) = L_d$.

$$L_z = \frac{\tau_i}{\tau_f}\left\{L_y\left(\sigma_{min2}\right) - \frac{1}{2}\left[L_d\left(\sigma_{max2}\right) + \left(1 - \frac{\tau_f}{\tau_i}\right)\xi\right.\right.$$
$$\left.\left. - \left[\frac{r_f}{2}\left(\frac{V_m E_m}{V_f E_c}\frac{\sigma}{\tau_i} - \frac{1}{\rho}\right) - \sqrt{\left(\frac{r_f}{2\rho}\right)^2 + \frac{r_f V_m E_m E_f}{E_c \tau_i^2}\Gamma_i}\right]\right]\right\} \quad (8.3)$$

8.2.2 Case 2

On complete unloading, the interface counter-slip length L_y is less than the interface debonding length L_d, that is, $L_y(\sigma_{min2}) < L_d$. The unloading interface counter-slip length L_y can be determined by Equation 8.2. On reloading to σ_{max2}, the interface new-slip length L_z is less than the interface debonding length L_d, that is, $L_z(\sigma_{max2}) < L_d$. The reloading interface new-slip length L_z is determined by Equation 8.3.

8.2.3 Case 3

When the interface completely debonds, on complete unloading, the interface counter-slip length L_y is less than half matrix crack spacing $L_c/2$, that is, $L_y(\sigma_{min2}) < L_c/2$.

$$L_y = \left(1 - \frac{\tau_f}{\tau_i}\right)\xi + \frac{r_f V_m E_m}{4 V_f E_c \tau_i}\left(\sigma_{max2} - \sigma\right) \qquad (8.4)$$

Upon reloading to σ_{max2}, the interface new-slip length L_z is less than the half matrix crack spacing $L_c/2$, that is, $L_z(\sigma_{max2}) < L_c/2$.

$$L_z = L_y\left(\sigma_{min2}\right) - \frac{r_f V_m E_m}{4 V_f E_c \tau_i}\left(\sigma_{max2} - \sigma\right) \qquad (8.5)$$

8.2.4 Case 4

When the interface complete debonds, upon unloading to the transition stress $\sigma_{tr\_fu}(\sigma_{tr\_fu} > \sigma_{min2})$, the interface counter-slip length L_y approaches half matrix crack spacing $L_c/2$, that is, $L_y(\sigma_{tr\_fu}) = L_c/2$. When $\sigma > \sigma_{tr\_fu}$, the unloading interface counter-slip length L_y is less than the half matrix crack spacing $L_c/2$, that is, $L_y(\sigma > \sigma_{tr\_fu}) < L_c/2$. The unloading interface counter-slip length L_y is determined by Equation 8.4. When $\sigma_{min2} < \sigma < \sigma_{tr\_fu}$, the unloading interface counter-slip occurs over the entire matrix crack spacing $L_c/2$, that is, $L_y(\sigma < \sigma_{tr\_fu}) = L_c/2$.

Upon reloading to the transition stress σ_{tr_fr} ($\sigma_{tr\_fr} < \sigma_{max2}$), the interface new-slip length L_z approaches half matrix crack spacing $L_c/2$. When $\sigma < \sigma_{tr\_fr}$, the interface new-slip length L_z is less than the half matrix crack spacing $L_c/2$, that is, $L_z(\sigma < \sigma_{tr\_fr}) < L_c/2$. The interface new-slip length L_z is determined by Equation (8.5). When $\sigma_{tr\_fr} < \sigma < \sigma_{max2}$, the interface new-slip length occurs over the entire matrix crack spacing $L_c/2$, that is, $L_z(\sigma > \sigma_{tr\_fr}) = L_c/2$.

8.2.5 Hysteresis Constitutive Relationship

The unloading and reloading stress-strain relationships for the fiber/matrix interface partial debonding and fiber partial sliding relative to the matrix in the interface debonding region are described using the following equations:

$$
\begin{aligned}
\varepsilon_{c\_pu} &= \frac{2\sigma L_d}{V_f E_f L_c} + \frac{2\tau_f}{r_f E_f L_c}\xi^2 + \frac{4\tau_f}{r_f E_f L_c}\xi\left(L_d-\xi\right) + \frac{4\tau_i}{r_f E_f L_c}\left(L_y-\xi\right)^2 \\
&\quad - \frac{2\tau_i}{r_f E_f L_c}\left(2L_y-\xi-L_d\right)^2 + \frac{2\sigma_{fo}}{E_f L_c}\left(\frac{L_c}{2}-L_d\right) \\
&\quad + \frac{2r_f}{\rho E_f l_c}\left[\frac{V_m}{V_f}\sigma_{mo} + \frac{2\tau_f}{r_f}\xi + \frac{2\tau_i}{r_f}\left(2L_y-\xi-L_d\right)\right] \\
&\quad \times\left[1-\exp\left(-\rho\frac{L_c/2-L_d}{r_f}\right)\right] - \left(\alpha_c-\alpha_f\right)\Delta T,
\end{aligned}
\tag{8.6}
$$

$$
\begin{aligned}
\varepsilon_{c\_pr} &= \frac{2\sigma}{V_f E_f L_c}L_d - \frac{4\tau_f}{r_f E_f L_c}L_z^2 + \frac{2\tau_f}{r_f E_f L_c}\left(2L_z-\xi\right)^2 - \frac{4\tau_f}{r_f E_f L_c}\left(2L_z-\xi\right)\left(L_d-\xi\right) \\
&\quad + \frac{4\tau_i}{r_f E_f L_c}\left(L_y-\xi\right)^2 - \frac{2\tau_i}{r_f E_f L_c}\left(2L_y-\xi-L_d\right)^2 + \frac{2\sigma_{fo}}{E_f L_c}\left(\frac{L_c}{2}-L_d\right) \\
&\quad + \frac{2r_f}{\rho E_f L_c}\left[\frac{V_m}{V_f}\sigma_{mo} - \frac{2\tau_f}{r_f}\left(2L_z-\xi\right) + \frac{2\tau_i}{r_f}\left(2L_y-\xi-L_d\right)\right] \\
&\quad \times\left[1-\exp\left(-\rho\frac{L_c/2-L_d}{r_f}\right)\right] - \left(\alpha_c-\alpha_f\right)\Delta T.
\end{aligned}
\tag{8.7}
$$

When the fiber complete slides relative to the matrix upon unloading and subsequent reloading, the unloading stress-strain relationship is divided into two parts. When $\sigma > \sigma_{tr\_pu}$, the unloading strain is given by Equation 8.6; when $\sigma < \sigma_{tr\_pu}$, the unloading strain is given by Equation 8.6 by setting $L_y = L_d$. The reloading stress-strain relationship is divided into two parts. When $\sigma < \sigma_{tr\_pr}$, the reloading strain is given by Equation 8.7; when $\sigma > \sigma_{tr\_pr}$, the reloading strain is given by Equation 8.7 by setting $L_z = L_d$.

The unloading and reloading stress-strain relationships for the fiber/matrix interface completely debonding and the fiber partially sliding relative to the matrix in the fiber/matrix interface debonding region are described using the following equation:

$$\varepsilon_{c\_fu} = \frac{\sigma}{V_f E_f} - \frac{2\tau_f}{r_f E_f L_c}\xi^2 + \frac{2\tau_f}{r_f E_f}\xi + \frac{4\tau_i}{r_f E_f L_c}\left(L_y - \xi\right)^2$$
$$- \frac{2\tau_i}{r_f E_f L_c}\left(2L_y - \xi - \frac{L_c}{2}\right)^2 - \left(\alpha_c - \alpha_f\right)\Delta T, \tag{8.8}$$

$$\varepsilon_{c\_fr} = \frac{\sigma}{V_f E_f} - \frac{4\tau_f}{r_f E_f L_c}L_z^2 + \frac{2\tau_f}{r_f E_f L_c}\left(2L_z - \xi\right)^2 - \frac{4\tau_f}{r_f E_f L_c}\left(2L_z - \xi\right)\left(L_y - \xi\right)$$
$$+ \frac{4\tau_i}{r_f E_f L_c}\left(L_y - \xi\right)^2 - \frac{4\tau_f}{r_f E_f L_c}\left(2L_z - \xi\right)\left(\frac{L_c}{2} - L_y\right)$$
$$- \frac{2\tau_i}{r_f E_f L_c}\left(2L_y - \xi - \frac{L_c}{2}\right)^2 - \left(\alpha_c - \alpha_f\right)\Delta T. \tag{8.9}$$

When the fiber complete slides relative to the matrix on unloading and subsequent reloading, the unloading stress-strain relationship is divided into two parts. When $\sigma > \sigma_{tr\_fu}$, the unloading strain is given by Equation 8.8; when $\sigma < \sigma_{tr\_fu}$, the unloading strain is given by Equation 8.8 by setting $L_y = L_c/2$. The reloading stress-strain relationship is also divided into two parts. When $\sigma < \sigma_{tr\_fr}$, the reloading strain is given by Equation 8.9; when $\sigma > \sigma_{tr\_fr}$, the reloading strain is given by Equation 8.9 by setting $L_z = L_c/2$.

8.3 EXPERIMENTAL COMPARISONS

Under cyclic loading/unloading, the loading directions were along the longitudinal fiber tow for 2D and 2.5D CMCs, the axial fibers at a small angle θ for 3D CMCs, and the longitudinal nonwoven cloth for 3D needled CMCs. In order to characterize the fiber architectures, an effective coefficient of the fiber volume content along the loading direction (ECFL) is defined using the following equation:

$$\phi = \frac{V_{f\_axial}}{V_f}, \tag{8.10}$$

where V_f and V_{f_axial} denote the total fiber volume fraction in the composite and the effective fiber volume fraction in the cyclic loading direction.

The values for parameter ϕ, which can be used to characterize fiber architectures for 2D, 2.5D, 3D, and needled CMCs are 0.5, 0.75, 0.93, and 0.375, respectively. The fatigue hysteresis loops of needled, 2D, 2.5D, and 3D C/SiC and 2D SiC/SiC composite under multiple fatigue loading are predicted, and the fiber/matrix interface slip lengths (i.e., interface counter-slip length and interface new-slip length) are analyzed.

8.3.1 C/SiC Composite

Mei and Cheng [33] investigated cyclic fatigue behavior of needled, 2D, 2.5D, and 3D C/SiC composite under multiple fatigue loading sequences. The fatigue hysteresis loops and the fiber/matrix interface slip length as a function of applied stress are predicted.

For a needled C/SiC composite, the experimental and predicted fatigue hysteresis loops under multiple fatigue peak stresses σ_{max1} = 95 MPa, σ_{max2} = 125 MPa, and σ_{max3} = 155 MPa are shown in Figure 8.1a, and the interface counter-slip length and interface new-slip length approach 64.8%, 62.1%, and 86.1% of the interface debonding length on unloading/reloading, as shown in Figure 8.1b.

For a 2D C/SiC composite, the experimental and predicted fatigue hysteresis loops under multiple fatigue peak stresses σ_{max1} = 180 MPa, σ_{max2} = 200 MPa, and σ_{max3} = 220 MPa are shown in Figure 8.2a, and the fiber/matrix interface counter-slip length and interface new-slip length approach the entire interface debonded length upon unloading/reloading, as shown in Figure 8.2b.

For a 2.5D C/SiC composite, the experimental and predicted fatigue hysteresis loops under multiple fatigue peak stresses σ_{max1} = 95 MPa and

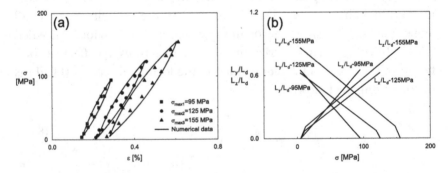

FIGURE 8.1 (a) Experimental and predicted hysteresis loops and (b) interface sliding ratio of a needled C/SiC composite under multiple loading stress.

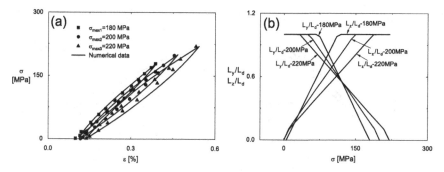

FIGURE 8.2   (a) Experimental and predicted hysteresis loops, and (b) interface sliding ratio of a 2D C/SiC composite under multiple loading stress.

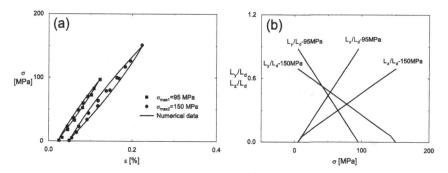

FIGURE 8.3   (a) Experimental and predicted hysteresis loops and (b) interface sliding ratio of a 2.5D C/SiC composite under multiple loading stress.

σ_{max2} = 150 MPa are shown in Figure 8.3a, and the interface counter-slip length and interface new-slip length approach 87.9% and 69%, respectively, of the interface debonding length on unloading/reloading, as shown in Figure 8.3b.

For a 3D C/SiC composite, the experimental and predicted fatigue hysteresis loops under multiple fatigue peak stresses σ_{max1} = 265 MPa, σ_{max2} = 285 MPa, and σ_{max3} = 300 MPa are shown in Figure 8.4a, and the fiber/matrix interface counter-slip length and interface new-slip length approach 63.9%, 64.9%, and 64.2%, respectively, of the interface debonding length on unloading/reloading, as shown in Figure 8.4b.

8.3.2 SiC/SiC Composite

McNulty and Zok [34] investigated the cyclic fatigue behavior of 2D SiC/SiC composite under multiple fatigue loading sequences.

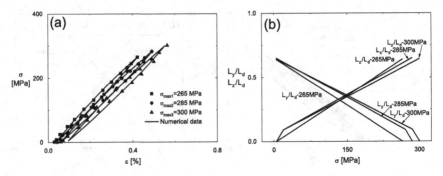

FIGURE 8.4 (a) Experimental and predicted hysteresis loops and (b) interface sliding ratio of a 3D C/SiC composite under multiple loading stress.

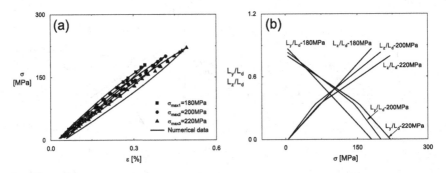

FIGURE 8.5 (a) Experimental and predicted hysteresis loops and (b) interface sliding ratio of a 2D SiC/SiC composite under multiple loading stress.

Experimental and predicted fatigue hysteresis loops under multiple fatigue peak stresses σ_{max1} = 180 MPa, σ_{max2} = 200 MPa, and σ_{max3} = 220 MPa are shown in Figure 8.5a, and the interface counter-slip length and interface new-slip length approach 86.6%, 83.1%, and 79.5%, respectively, of the interface debonding length, as shown in Figure 8.5b.

8.4 DISCUSSION

The damage evolution of fiber-reinforced CMCs under multiple fatigue loading sequences may be affected by the fiber volume fraction, matrix crack spacing, fatigue peak stress, and fatigue stress range. Effects of these factors are investigated to establish the relationships between fatigue hysteresis and internal damage inside fiber-reinforced CMCs.

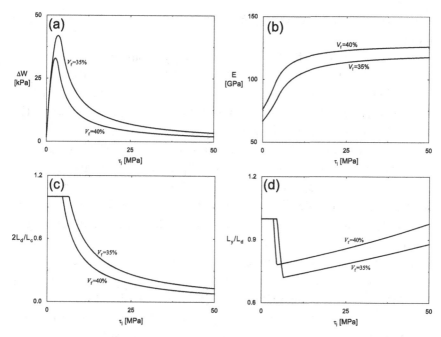

FIGURE 8.6   Effect of fiber volume content (i.e., V_f = 35 and 40%) on (a) fatigue hysteresis dissipated energy, (b) fatigue hysteresis modulus, (c) interface debonding length, and (d) interface counter-slip length under multiple loading fatigue peak stress levels σ_{max1} = 100 MPa and σ_{max2} = 180 MPa with τ_f = 1MPa.

8.4.1 Effect of Fiber Volume Content

Effect of fiber volume content (i.e., V_f = 35 and 40%) on the fatigue hysteresis energy dissipation, fatigue hysteresis modulus, interface debonding length, and interface counter-slip length under multiple fatigue peak stress levels σ_{max1} = 100 MPa and σ_{max2} = 180 MPa are shown in Figure 8.6.

When V_f = 35%, the fatigue hysteresis energy dissipation increases from ΔW = 3.4 kPa at τ_i = 50 MPa to ΔW = 26.4 kPa at τ_i = 6.5 MPa, and the fatigue hysteresis modulus decreases from E = 117 GPa at τ_i = 50 MPa to E = 94.3 GPa at τ_i = 6.5 MPa, corresponding to the interface slip Case 2; the fatigue hysteresis dissipated energy increases to ΔW = 38.1 kPa at τ_i = 4.5 MPa, the fatigue hysteresis modulus decreases to E = 85.7 GPa at τ_i = 4.5 MPa, corresponding to the interface slip Case 3; the fatigue hysteresis dissipated energy increases to the peak value ΔW = 41.8 kPa at τ_i = 3.5 MPa and decreases to ΔW = 22 kPa at τ_i = 1 MPa; and the fatigue hysteresis modulus decreases to $E = 70.1$ GPa, corresponding to the interface slip Case 4.

When $V_f = 40\%$, the fatigue hysteresis energy dissipation increases from $\Delta W = 2.1$ kPa at $\tau_i = 50$ MPa to $\Delta W = 22.9$ kPa at $\tau_i = 4.5$ MPa, and the fatigue hysteresis modulus decreases from $E = 126$ GPa at $\tau_i = 50$ MPa to $E = 101$ GPa at $\tau_i = 4.5$ MPa, corresponding to the interface slip Case 2; the fatigue hysteresis dissipated energy increases to $\Delta W = 29.5$ kPa at $\tau_i = 3.5$ MPa, and the fatigue hysteresis modulus decreases to $E = 95.5$ GPa at $\tau_i = 3.5$ MPa, corresponding to the interface slip Case 3; the fatigue hysteresis dissipated energy increases to the peak value $\Delta W = 32.9$ kPa at $\tau_i = 2.5$ MPa and decreases to $\Delta W = 20.5$ kPa at $\tau_i = 1$ MPa; and the fatigue hysteresis modulus decreases to $E = 80.7$ GPa at $\tau_i = 2.5$ MPa, corresponding to the interface slip Case 4.

With increasing fiber volume content, the interface debonding length and interface slip lengths, that is, interface counter-slip length and interface new-slip length, decrease when the interface partial and complete debond; the fatigue hysteresis energy dissipation decreases for the interface slip Cases 1, 2, and 3; however, when the interface slip corresponds to the Case 4, the fatigue hysteresis energy dissipation approaches the same value, and the fatigue hysteresis modulus increases.

When the fiber volume content increases, the range and extent of fiber/matrix interface frictional slip between the fiber and the matrix in the interface debonding region decrease when the interface slip corresponds to Cases 1, 2, and 3, leading to the decreasing fatigue hysteresis energy dissipation and the increasing fatigue hysteresis modulus.

8.4.2 Effect of Matrix Crack Spacing

Effect of matrix crack spacing (i.e., $L_c = 20r_f$ and $30r_f$) on the fatigue hysteresis energy dissipation, fatigue hysteresis modulus, interface debonding length and interface counter-slip length under multiple fatigue peak stress levels $\sigma_{max1} = 100$ MPa and $\sigma_{max2} = 180$ MPa are shown in Figure 8.7.

When $L_c = 20r_f$, the fatigue hysteresis energy dissipation increases from $\Delta W = 4.6$ kPa at $\tau_i = 50$ MPa to $\Delta W = 25.9$ kPa at $\tau_i = 8.8$ MPa, and the fatigue hysteresis modulus decreases from $E = 115$ GPa at $\tau_i = 50$ MPa to $E = 94$ GPa at $\tau_i = 8.8$ MPa, corresponding to the interface slip Case 2; the fatigue hysteresis dissipated energy increases to $\Delta W = 37.2$ kPa at $\tau_i = 6.1$ MPa, and the fatigue hysteresis modulus decreases to $E = 85$ GPa at $\tau_i = 6.1$ MPa, corresponding to the interface slip Case 3; the fatigue hysteresis

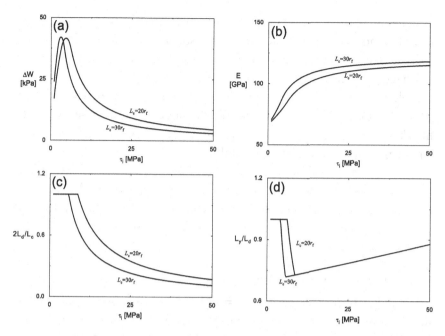

FIGURE 8.7   Effect of matrix crack spacing (i.e., $L_c = 20r_f$ and $30r_f$) on (a) fatigue hysteresis dissipated energy, (b) fatigue hysteresis modulus, (c) interface debonding length, and (d) interface counter-slip length under multiple loading fatigue peak stress levels $\sigma_{max1} = 100$ MPa and $\sigma_{max2} = 180$ MPa with $\tau_f = 1$MPa.

dissipated energy increases to the peak value $\Delta W = 41.5$ kPa at $\tau_i = 4.4$ MPa and decreases to $\Delta W = 17.1$ kPa at $\tau_i = 1$ MPa; and the fatigue hysteresis modulus decreases to $E = 69$ GPa at $\tau_i = 1$ MPa, corresponding to the interface slip Case 4.

When $L_c = 30r_f$, the fatigue hysteresis dissipated energy increases from $\Delta W = 3.1$ kPa at $\tau_i = 50$ MPa to $\Delta W = 25.8$ kPa at $\tau_i = 5.9$ MPa, and the fatigue hysteresis modulus decreases from $E = 118$ GPa at $\tau_i = 50$ MPa to $E = 95$ GPa at $\tau_i = 5.9$ MPa, corresponding to the interface slip Case 2; the fatigue hysteresis dissipated energy increases to $\Delta W = 38.6$ kPa at $\tau_i = 3.9$ MPa, and the fatigue hysteresis modulus decreases to $E = 85$ GPa at $\tau_i = 3.9$ MPa, corresponding to the interface slip Case 3; the fatigue hysteresis dissipated energy increases to the peak value $\Delta W = 42$ kPa at $\tau_i = 3$ MPa and decreases to $\Delta W = 24$ kPa at $\tau_i = 1$ MPa; and the fatigue hysteresis modulus decreases to $E = 70.6$ GPa at $\tau_i = 1$ MPa, corresponding to the interface slip Case 4.

With increasing matrix crack spacing, the extent of fiber/matrix interface debonding and interface frictional slip in the entire matrix crack spacing decrease at the same interface shear stress when the fiber/matrix interface partially debonded and increase when the fiber/matrix interface completely debonded; the fatigue hysteresis dissipated energy decreases when the fiber/matrix interface partial debonded and increases when the fiber/matrix interface complete debonded at the same interface shear stress; and the fatigue hysteresis modulus increases.

When matrix crack spacing increases, the extent of fiber/matrix interface frictional slip between the fiber and the matrix in the interface debonded region is affected by the interface debonding, that is, decrease when the interface partially debonds and increase when the interface completely debonds.

8.4.3 Effect of Low Peak Stress Level

Effect of low peak stress level σ_{max1} = 100 and 140 MPa on the fatigue hysteresis dissipated energy, fatigue hysteresis modulus, fiber/matrix interface debonded length, and interface counter-slip length under high fatigue peak stress level σ_{max2} = 180 MPa are shown in Figure 8.8.

When σ_{max1} = 100 MPa, the fatigue hysteresis dissipated energy increases from ΔW = 7.3 kPa at τ_i = 50 MPa to ΔW = 22.9 kPa at τ_i = 15.2 MPa, and the fatigue hysteresis modulus decreases from E = 99.8 GPa at τ_i = 50 MPa to E = 87.7 GPa at τ_i = 15.2 MPa, corresponding to the interface slip Case 2; the fatigue hysteresis dissipated energy increases to ΔW = 31.6kPa at τ_i = 11 MPa, and the fatigue hysteresis modulus decreases to E = 82.2 GPa at τ_i = 11 MPa, corresponding to the interface slip Case 3; the fatigue hysteresis dissipated energy increases to the peak value ΔW = 35.5 kPa at τ_i = 8 MPa and then decreases to ΔW = 11.4 kPa at τ_i = 1 MPa, and the fatigue hysteresis modulus decreases to E = 68.2 GPa at τ_i = 1 MPa, corresponding to the interface slip Case 4.

When σ_{max1} = 140 MPa, the fatigue hysteresis dissipated energy increases from ΔW = 8.6 kPa at τ_i = 50 MPa to ΔW = 19.3 kPa at τ_i = 19.5 MPa and the fatigue hysteresis modulus decreases from E = 88.1 GPa at τ_i = 50 MPa to E = 81.5 GPa at τ_i = 19.5 MPa, corresponding to the interface slip Case 2; the fatigue hysteresis dissipated energy increases to ΔW = 25.7 kPa at τ_i = 14.2 MPa, and the fatigue hysteresis modulus decreases to E = 78.1 GPa at τ_i = 14.2 MPa, corresponding to the interface slip Case 3; the fatigue hysteresis dissipated energy increases to the peak value ΔW = 28.8 kPa at

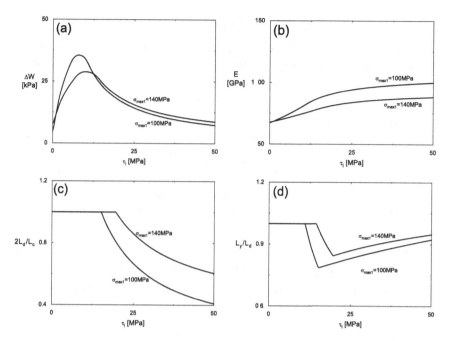

FIGURE 8.8   Effect of low fatigue peak stress (i.e., σ_{max_1} = 100 and 140 MPa) on (a) fatigue hysteresis dissipated energy, (b) fatigue hysteresis modulus, (c) interface debonding length, and (d) interface counter slip length under high fatigue peak stress level σ_{max2} = 180 MPa with τ_f = 1MPa.

τ_i = 10 MPa and then decreases to ΔW = 11.4 kPa at τ_i = 1 Mpa; and the fatigue hysteresis modulus decreases to E = 68.2 GPa at τ_i = 1 MPa, corresponding to the interface slip Case 4.

With increasing low peak stress, the fiber/matrix interface debonded length and interface slip lengths, that is, interface counter-slip length and interface new-slip length, under high peak stress increase when the fiber/matrix interface partially debonded; the fatigue hysteresis dissipated energy increases for the fiber/matrix interface slip Cases 1, 2 and 3; however, when the interface slip corresponds to the Case 4, the fatigue hysteresis dissipated energy decreases first and then approaches the same value when the interface shear stress decreases to τ_i = 1 MPa, and the fatigue hysteresis modulus decreases.

When the low peak stress level increases, the range and the extent of fiber/matrix interface frictional slip between the fiber and the matrix in the interface debonded region increase when the interface slip corresponds to the Cases 1, 2, and 3.

8.4.4 Effect of High Peak Stress Level

The effect of high peak stress level σ_{max2} = 140 and 180 MPa on the fatigue hysteresis dissipated energy, fatigue hysteresis modulus, fiber/matrix interface debonded length, and interface counter-slip length at low fatigue peak stress level σ_{max1} = 120 MPa are shown in Figure 8.9.

When σ_{max2} = 140 MPa, the fatigue hysteresis dissipated energy increases from ΔW = 1.3 kPa at τ_i = 50 MPa to ΔW = 1.9 kPa at τ_i = 41.5 MPa, and the fatigue hysteresis modulus decreases from E = 118.4 GPa at τ_i = 50 MPa to E = 117.9 GPa at τ_i = 41.5 MPa, corresponding to the interface slip Case 1; the fatigue hysteresis dissipated energy increases to ΔW = 17.9 kPa at τ_i = 4.5 MPa, and the fatigue hysteresis modulus decreases to E = 91.8 GPa at τ_i = 4.5 MPa, corresponding to the interface slip Case 2; the fatigue hysteresis dissipated energy increases to ΔW = 23 kPa at τ_i = 3.5 MPa, and the fatigue hysteresis modulus decreases to E = 85.7 GPa at τ_i = 3.5 MPa, corresponding to the interface slip Case 3; the fatigue hysteresis dissipated

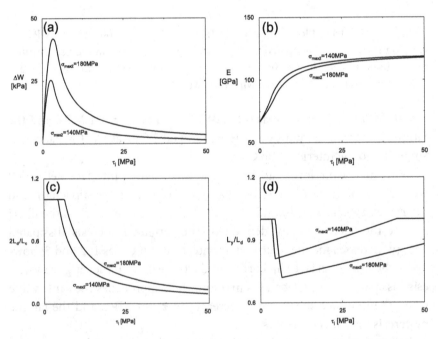

FIGURE 8.9 Effect of high fatigue peak stress (i.e., σ_{max2} = 140 and 180 MPa) on (a) fatigue hysteresis dissipated energy, (b) fatigue hysteresis modulus, (c) interface debonding length, and (d) interface counter slip length with low fatigue peak stress level σ_{max1} = 120 MPa and τ_f = 1 MPa.

energy increases to the peak value ΔW = 25.3 kPa at τ_i = 2.5 MPa, and decreases to ΔW = 16 kPa at τ_i = 1 MPa, and the fatigue hysteresis modulus decreases to E = 71.2 GPa, corresponding to the interface slip Case 4.

When σ_{max2} = 180 MPa, the fatigue hysteresis dissipated energy increases from ΔW = 3.4 kPa at τ_i = 50 MPa to ΔW = 26.4 kPa at τ_i = 6.5 MPa, and the fatigue hysteresis modulus decreases from E = 117.7 GPa at τ_i = 50 MPa to E = 94.3 GPa at τ_i = 6.5 MPa, corresponding to the interface slip Case 2; the fatigue hysteresis dissipated energy increases to ΔW = 38.1 kPa at τ_i = 4.5 MPa, and the fatigue hysteresis modulus decreases to E = 85.7 GPa at τ_i = 4.5 MPa, corresponding to the interface slip Case 3; and the fatigue hysteresis dissipated energy increases to the peak value ΔW = 41.8 kPa at τ_i = 3.5 MPa and decreases to ΔW = 22 kPa at τ_i = 1 MPa, and the fatigue hysteresis modulus decreases to E = 70.1 GPa at τ_i = 1 MPa, corresponding to the interface slip Case 4.

With increasing high peak stress, the fiber/matrix interface debonded length increases when the interface partially debonded; the interface slip lengths, that is, interface counter-slip length and interface new-slip length, decreases when the interface partially debonded due to the longer interface debonded length; the fatigue hysteresis dissipated energy increases for the interface slip Cases 1, 2, 3, and 4; and the fatigue hysteresis modulus decreases.

When high peak stress increases, the fiber/matrix interface debonded length increases; the range and the extent of the interface frictional slip between the fiber and the matrix in the interface debonded region also increase, leading to the increase of fatigue hysteresis energy dissipation at the same interface shear stress when the interface slip corresponds to the Cases 1, 2, 3, and 4.

8.4.5 Effect of Fatigue Stress Range

The effect of stress range (i.e., $\Delta\sigma$ = 100 and 120 MPa) on the fatigue hysteresis dissipated energy, fatigue hysteresis modulus, fiber/matrix interface debonded length, and interface counter-slip length under multiple fatigue peak stress levels σ_{max1} = 100 MPa and σ_{max2} = 180 MPa are shown in Figure 8.10.

When $\Delta\sigma$ = 100 MPa, the fatigue hysteresis dissipated energy increases from ΔW = 0.59 kPa at τ_i = 50 MPa to ΔW = 4.4 kPa at τ_i = 6.6 MPa, and the fatigue hysteresis modulus decreases from E = 119.7 GPa at τ_i = 50 MPa to E = 105.4 GPa at τ_i = 6.6 MPa, corresponding to the interface slip Case 2;

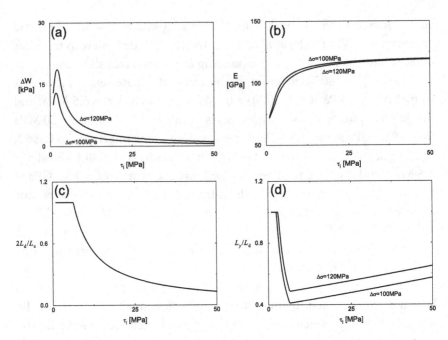

FIGURE 8.10 Effect of fatigue stress range (i.e., $\Delta\sigma$ = 100 and 120 MPa) on (a) fatigue hysteresis dissipated energy, (b) fatigue hysteresis modulus, (c) interface debonding length, and (d) interface counter-slip length under multiple loading fatigue peak stress levels σ_{max1} = 100 MPa and σ_{max2} = 180 MPa with τ_f = 1MPa.

the fatigue hysteresis dissipated energy increases to ΔW = 11.8kPa at τ_i = 2.5 MPa, and the fatigue hysteresis modulus decreases to E = 85.6 GPa at τ_i = 2.5 MPa, corresponding to the interface slip Case 3; and the fatigue hysteresis dissipated energy increases to the peak value ΔW = 12.9 kPa at τ_i = 2 MPa and decreases to ΔW = 10.2 kPa at τ_i = 1 MPa, and the fatigue hysteresis modulus decreases to E = 73.3 GPa at τ_i = 1 MPa, corresponding to the interface slip Case 4.

When $\Delta\sigma$ = 120 MPa, the fatigue hysteresis dissipated energy increases from ΔW = 1 kPa at τ_i = 50 MPa to ΔW = 7.6 kPa at τ_i = 6.6 MPa, and the fatigue hysteresis modulus decreases from E = 119.2 GPa at τ_i = 50 MPa to E = 102.5 GPa at τ_i = 6.6 MPa, corresponding to the interface slip Case 2; the fatigue hysteresis dissipated energy increases to ΔW = 17.1 kPa at τ_i = 2.9 MPa, and the fatigue hysteresis modulus decreases to E = 85.5 GPa at τ_i = 2.9 MPa, corresponding to the interface slip Case 3; and the fatigue hysteresis dissipated energy increases to the peak value ΔW = 18.6 kPa at

τ_i = 2.2 MPa and decreases to ΔW = 13 kPa at τ_i = 1 MPa, and the fatigue hysteresis modulus decreases to E = 72.1 GPa at τ_i = 1 MPa, corresponding to the interface slip Case 4.

With increasing stress range, the fiber/matrix interface slip lengths, that is, interface counter-slip length and interface new-slip length increase when the fiber/matrix interface partially and completely debonded; however, the fiber/matrix interface debonded length depends on the peak stress, which is not affected by stress range; the fatigue hysteresis dissipated energy increases due to the increase of extent and range of interface frictional slip between the fiber and the matrix; and the fatigue hysteresis modulus decreases.

8.5 SUMMARY AND CONCLUSION

In this chapter, mechanical hysteresis behavior for fiber-reinforced CMCs subjected to multiple fatigue loading sequences was investigated. The fatigue hysteresis energy dissipation and fiber/matrix interface slip corresponding to different fiber volume content, matrix crack spacing, low and high-stress levels, and fatigue stress range were investigated. The fatigue hysteresis loops of needled, 2D, 2.5D, and 3D C/SiC and 2D SiC/SiC composites under multiple loading stress levels were predicted.

REFERENCES

1. Pryce AW, Smith PA. Matrix cracking in unidirectional ceramic matrix composites under quasi-static and cyclic loading. *Acta Metall. Mater.* 1993; 41:1269–1281.
2. Askarinejad S, Rahbar N, Sabelkin V, Mall S. Mechanical behavior of a notched oxide/oxide ceramic matrix composite in combustion environment: Experiments and simulations. *Compos. Struct.* 2015; 127:77–86.
3. Gowayed Y, Abouzeida E, Smyth I, Ojard G, Ahmad J, Santhosh U, Jefferson G. The role of oxidation in time-dependent response of ceramic-matrix composites. *Compos. Part B Eng.* 2015; 76:20–30.
4. Sabelkin V, Zawada L, Mall S. Effects of combustion and salt-fog exposure on fatigue behavior of two ceramic matrix composites and a superalloy. *J. Mater. Sci.* 2015; 50:5204–5213.
5. Dong N, Zuo X, Liu Y, Zhang L, Cheng L. 2016. Fatigue behavior of 2D C/SiC composites modified with Si-B-C ceramic in static air. *J. Euro. Ceram. Soc.* 2016; 36:3691–3696.
6. Luo Z, Cao H, Ren H, Zhou X. 2016. Tension-tension fatigue behavior of a PIP SiC/SiC composite at elevated temperature in air. *Ceram. Int.* 2016; 42:3250–3260.

7. Ruggles-Wrenn MB, Lee MD. Fatigue behavior of an advanced SiC/SiC ceramic composite with a self-healing matrix at 1300°C in air and in steam. *Mater. Sci. Eng. A* 2016; 677:438–445.
8. Wooh SC, Daniel IM. Real-time ultrasonic monitoring for fiber-matrix debonding in ceramic-matrix composite. *Mech. Mater.* 1994; 17:379–388.
9. Loutas TH, Kostopoulos V. Health monitoring of carbon/carbon, woven reinforced composites. Damage assessment by using advanced signal processing techniques. Part I: Acoustic emission monitoring and damage mechanisms evolution. *Compos. Sci. Technol.* 2009; 69:265–272.
10. Momon S, Moevus M, Godin N, R'Mili M, Reynaud P, Fantozzi G, Fayolle G. Acoustic emission and lifetime prediction during static fatigue tests on ceramic matrix composite at high temperature under air. *Compos. Part A* 2010; 41:913–918.
11. Whitlow T, Jones E, Przybyla C. In-situ damage monitoring of a SiC/SiC ceramic matrix composite using acoustic emission and digital image correlation. *Compos. Struct.* 2016; 158:245–251.
12. Simon C, Rebillat F, Camus G. Electrical resistivity monitoring of a SiC/[Si-B-C] composite under oxidizing environments. *Acta Mater.* 2017; 132:586–597.
13. Vagaggini E, Domergue JM, Evans AG. Relationships between hysteresis measurements and the constituent properties of ceramic matrix composites: I, Theory. *J. Am. Ceram. Soc.* 1995; 78:2709–2720.
14. Domergue JM, Vagaggini E, Evans AG. Relationships between hysteresis measurements and the constituent properties of ceramic matrix composites: II, Experimental studies on unidirectional materials. *J. Am. Ceram. Soc.* 1995; 78:2721–2731.
15. Solti JP, Mall S, Robertson DD. Modeling damage in unidirectional ceramic-matrix composites. *Compos. Sci. Technol.* 1995; 54:55–66.
16. Reynaud P. Cyclic fatigue of ceramic-matrix composites at ambient and elevated temperatures. *Compos. Sci. Technol.* 1996; 56:809–814.
17. Campbell CX, Jenkins MG. Use of unload/reload methodologies to investigate the thermal degradation of an alumina fiber-reinforced ceramic matrix composites. *Symposium on Environmental, Mechanical, and Thermal Properties and Performance of Continuous Fiber Ceramic Composite (CFCC) Materials and Components*, Seattle, Washington, May 18, 1999.
18. Fantozzi G, Reynaud P. Mechanical hysteresis in ceramic matrix composites. *Mater. Sci. Eng. A* 2009; 521–522, 18–23.
19. Li LB. Synergistic effects of fiber debonding and fracture on matrix cracking in fiber-reinforced ceramic-matrix composites. *Mater. Sci. Eng. A* 2017; 682:482–490.
20. Li LB. Modeling matrix cracking of fiber-reinforced ceramic-matrix composite under oxidation environment at elevated temperature. *Theor. Appl. Fract. Mech.* 2017; 87:110–119.
21. Li LB. Synergistic effects of stress-rupture and cyclic loading on strain response of fiber-reinforced ceramic-matrix composites at elevated temperature in oxidizing atmosphere. *Mater.* 2017; 10:182.

22. Li LB. Comparisons of damage evolution between 2D C/SiC and SiC/SiC ceramic-matrix composites under tension-tension cyclic fatigue at room and elevated temperatures. *Mater.* 2016; 9:844.

23. Li LB. Interface debonding and slipping of carbon fiber-reinforced ceramic-matrix composites under two-stage cyclic loading. *Compos. Interface* 2017; 24:417–445.

24. Li LB. Fatigue hysteresis behavior in fiber-reinforced ceramic-matrix composites at room and elevated temperatures. *Ceram. Int.* 2017; 43:2514–2624.

25. Lazzarin P, Berto F, Zappalorto M. Rapid calculation of notch stress intensity factors based on averaged strain energy density from coarse meshes: Theoretical bases and applications. *Int. J. Fatigue* 2010; 32:1559–1567.

26. Lazzarin P, Afshar R, Berto F. Notch stress internsity factors of flat plates with periodic sharp notches by using the strain energy density. *Theor. Appl. Fract. Mech.* 2012; 60:38–50.

27. Berto F, Lazzarin P, Marango C. Fatigue strength of notched specimens made of 40CrMoV13.9 under multiaxial loading. *Mater. Design* 2014; 54:57–66.

28. Li LB. Modeling of fatigue hysteresis behavior in carbon fiber-reinforced ceramic-matrix composite under multiple loading stress levels. *J. Compos. Mater.* 2017; 51:971–983.

29. Li LB. A hysteresis dissipated energy-based parameter for damage monitoring of carbon fiber-reinforced ceramic-matrix composites under fatigue loading. *Mater. Sci. Eng. A* 2015; 634:188–201.

30. Li LB. A hysteresis dissipated energy-based damage parameter for life prediction of carbon fiber-reinforced ceramic-matrix composites under fatigue loading. *Compos. Part B Eng.* 2015; 82:108–128.

31. Li LB. Damage development in fiber-reinforced ceramic-matrix composites under cyclic fatigue loading using hysteresis loops at room and elevated temperatures. *Int. J. Fract.* 2016; 199:39–58.

32. Li LB. A hysteresis energy dissipation based model for multiple loading damage in continuous fiber-reinforced ceramic-matrix composites. *Compos Part B* 2019; 162:259–273.

33. Mei H, Cheng L. Comparison of the mechanical hysteresis of carbon/ceramic-matrix composites with different fiber preforms. *Carbon* 2009; 47:1034–1042.

34. McNulty JC, Zok FW. Low-cycle fatigue of Nicalon-fiber-reinforced ceramic composites. *Compos. Sci. Technol.* 1999; 59:1597–1607.

Index

[Page numbers in *italics* denote figures; those in **bold** denote tables]

217

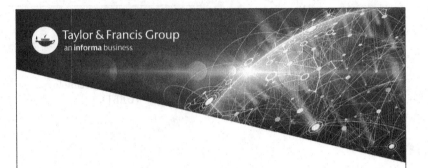

Printed in the United States
by Baker & Taylor Publisher Services